光明城

CITÉ LUCE

看见我们的未来

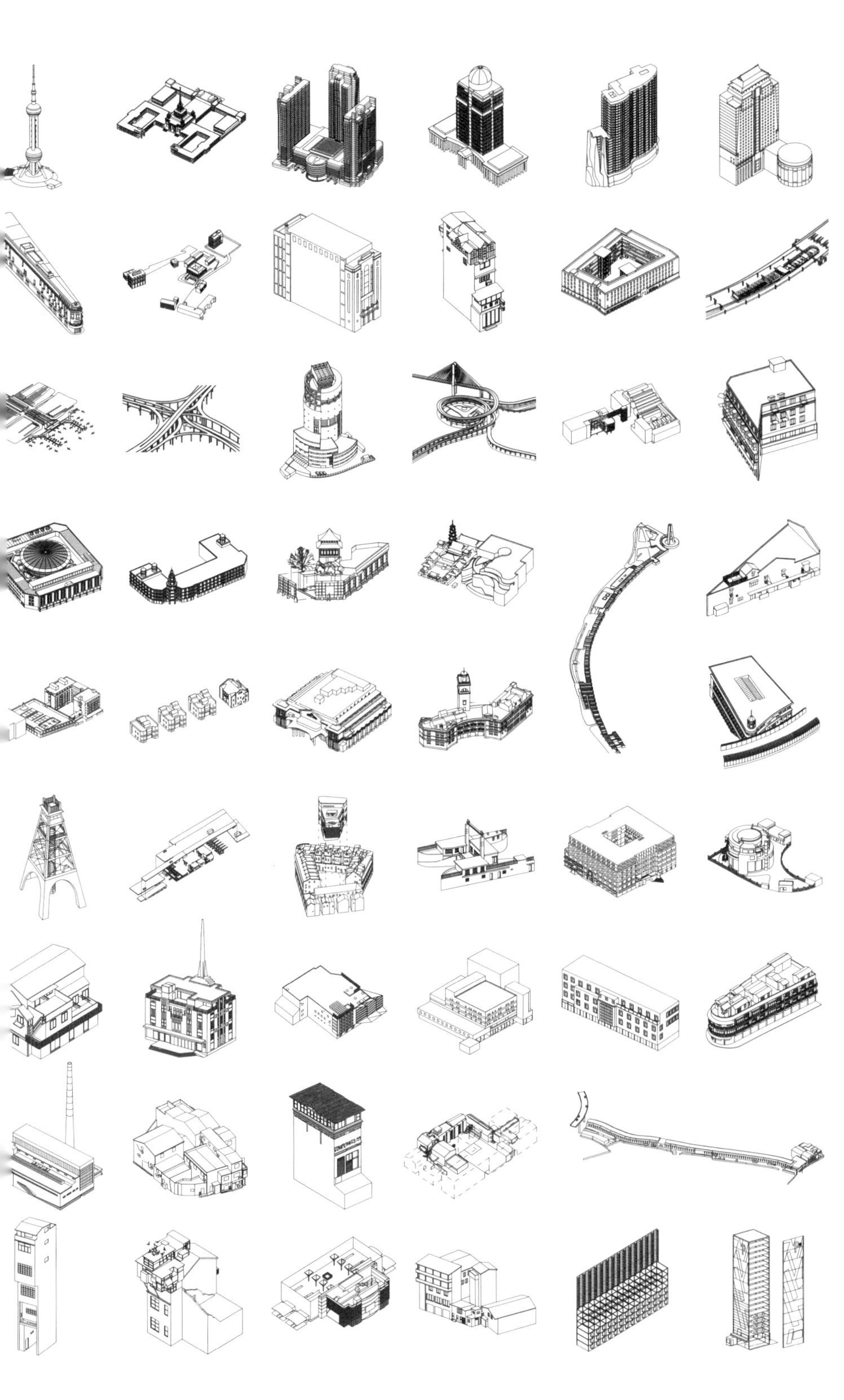

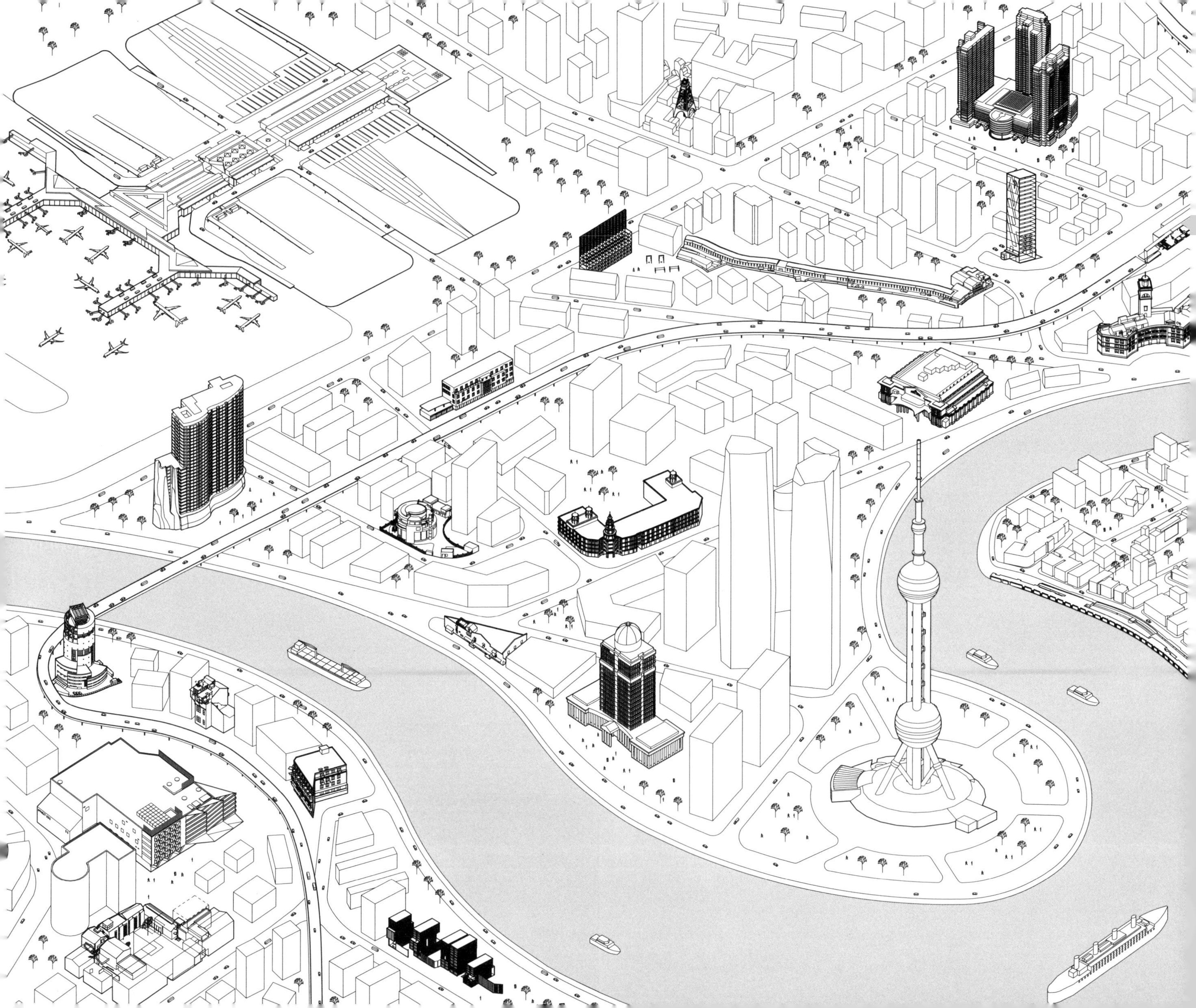

上海制造
MADE IN SHANGHAI

李翔宁　李丹锋　江嘉玮　著
Li Xiangning　Li Danfeng　Jiang Jiawei
合作者　塚本由晴
In collaboration with Yoshiharu Tsukamoto

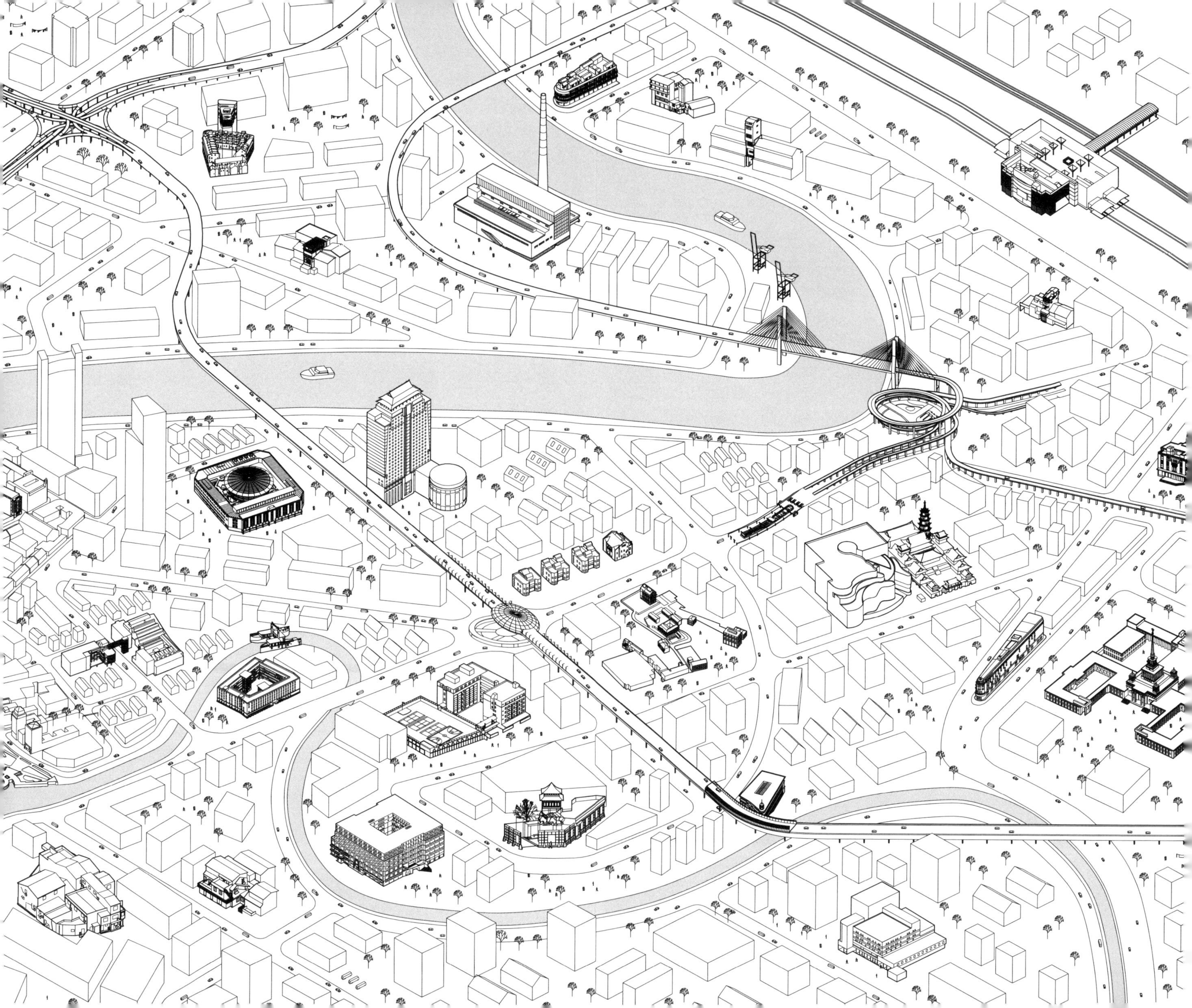

序一：让我们来读一本“魔都”指南

Preface I: A Guide to the "Devilish City"

朱大可
Zhu Dake

同济大学人文学院教授
Professor, School of Humanities, Tongji University

李翔宁等人著的《上海制造》，是城市景观织体的一个远东样本，汇集了上海“非主流建筑美学”的奇观式建筑，其中囊括了城市地标（东方明珠塔）、城市传奇（延安路立交桥龙柱）、城市文脉（大世界和上海一九叁三）、城市违章建筑（鸽舍）和贫民窟（虹镇老街）等各类货色。它是一个杂耍式拼贴的结果，向我们展示出“魔都”景观的奇异特性。

导致都市景观奇异化的原因是多样的，它不仅源于权力意志，炫耀政绩的意志，也缘于权力与资本及宗教结盟，针对空间逼仄的扩张策略，多重场所叠加而引发的空间畸变等。“奇异”是反常发育引发的效应，它不是城市管理的结果，更与城市浪漫主义无关。但它竟然改造了平庸的规划和设计理性，令城市散发出一种畸形、病态和自我分裂的气味。

法国文化批评家居伊·德波（Guy Debord）在其代表作《景观社会》中宣称，世界完成了从“商品的堆积”到“景观的庞大堆聚”的转变，而景观已成为“媒介时代”的本质。哪里有独立的表象，景观就会在哪里重构自己的法则。德波进而指出，明星、休闲旅游和城市化成为当今世界最突出的景观。在我看来，本书提供的全部例证，戏剧性地回应了这些跨时代的论断。

作为一部阐释性的建筑地图，本书试图导引文化旅行者（游客）去观看那些奇异的城市建筑，并对它们的功能和意义进行扼要评述，其中既有辛辣直白的批评，亦有暧昧的中性解读，也有爱不释之的褒扬。对于一个文化旅行者而言，没有比这更有趣的指南了。上海是一个巨大的地理迷津，它需要一本不同寻常的价值指南，以阐释那些有名和无名的奇异建筑，为它们的存在寻找理由。

耗资 27 亿元建造的平安金融大厦，一个罗马万神庙和现代矩形大厦的混合物，被一大堆的罗马柱所环绕，在其顶部安置一个体量过小的瓜皮帽式的穹顶，是神学景观和资本景观的一次低劣拼贴，成为上海式媚俗的一个典范之作。本书的解读，揭示了此类建筑的文化本性。

这种古怪的拼贴，正是上海建筑景观的一个基本特征。位于延安路高架与南北高架交接处的“龙柱”，是钢筋混凝土立柱和龙纹浮雕的古怪拼贴，常德路的“山景房”，用大型假山实体覆盖在公寓楼的侧面，形成恶俗的虚构性拼贴；而土黄色的古庙静安寺，则跟一侧的巨幅身体广告牌，形成了更具反讽性的拼贴。正是这种畸形的视觉组合，制造了类似“迪士尼乐园”的卡通效应。

但有一种拼贴似乎是我们能够容忍并予以期待的，它属于时间而非空间。延安东路和西藏南路交界处的“大世界”，一个旧历史和新时代的拼贴，至今依然矗立在迅速变化的都市中心。它因权利归属问题而被长期悬置，找不到重启的契机，却成为一个殖民地时期的建筑神话，与娱乐、流氓、戏子和妓女的传奇相混合，犹如一张立体的月份牌绘画，镶嵌于都市的深处，向我们提示关于殖民地的历史记忆。正如书中所描述的，这座上海昔日最大的室内游乐场，以游艺、杂耍和南北戏剧、曲艺为特色，是最具代表性的娱乐建筑，并最终成为上海景观的一个隐秘象征——原有的娱乐功能投射到整座城市，令其到处都散发着戏谑化、稚拙化和丑角化的文化气味。

另一个相似的例证，是位于虹口区沙泾路上的“上海一九叁三”。这座造型诡异的建筑，曾经是远东最大的屠宰场，“神秘幽深的光线”照亮了“错综复杂的走道”，在被钢和玻璃进行现代性改造之后，它保留了残留在墙体上的各种岁月印痕，犹如那些牲畜亡灵的喘息。华丽的表皮装饰，却难以掩饰其哥特式的阴郁风格。这是一种恒久的气息，深刻地渗入建筑物的骨架。而就在不远处，一座砖混结构的鸽舍，出现于老房子的顶部。屋主为每只鸽子搭建了专门的场所。但鸽舍所在的场地属于政府拆迁范围，估计该房彻底消失的日子不会太远。这是鸽舍、养鸽者和鸽子的共同命运。但它跟其他“违章搭建”的民居一起，成了都市景观中最卑贱、丑陋而富有生气的部分。

迄今为止，我还没有读到过此类反面的建筑导读，它的手指，不但指向那些奇异的地标，而且指向那些反常、畸形和速朽的事物，甚至指向被城市管理者所斥责的部分。但正是这些在逼仄空间里营造出来的怪屋，喊出了这座城市的居住本质。尽管城市的容量在急剧扩张，但市中心的空间张力，却仍在持续地强化之中。它逼迫怪屋生长，犹如怒放在行政官员梦境里的“恶之花”。对这些异化的建筑体的描述，与其说是一场充满戏谑性的导游，不如说是一些事先张扬的悼词，宣告着它们在未来的死亡，进而以平面印刷物的形态，预藏起这些古怪的都市遗产。

上海之所以被网民叫做“魔都”，无疑是基于它的城市景观特色。上海建筑具有中国任何城市都缺乏的魔幻性。只要你站在一个合适的地点加以观察，就不难发现，在阳光和雾霾的共同笼罩下，以黄浦江为中轴线，大多数建筑的外立面造型，都露出诡异神秘的哥特式面容。浦东完全继承了浦西的这种气质，它扩展了魔都的领地，将其径直推向大陆与海洋的分界线。“魔都”的魔幻风格，就是上海的魅力所在。它发出了女妖塞壬般的歌声，不倦地引诱着后现代游客的脆弱灵魂。

浦东张江
2013 年 11 月 28 日

Made in Shanghai, under the general editorship of Li Xiangning, is a sample of urban landscape in the Far East. It collects grotesque buildings of "non-mainstream architectural aesthetics" in Shanghai, including city landmarks (the Oriental Pearl TV Tower), city legends (Dragon-pillared overpass on Yan'an Road), cultural sites (the Great World and Shanghai 1933), the unapproved construction projects (pigeon houses) and slums (Hongzhen Old Street) and others. It is a vaudeville-style collage, which presents the uniqueness of the landscape of a "devilish city".

There are many reasons that contribute to a bizarre city landscape. It derives not only from the will of those in power to parade its political achievements, but also from an alliance between power and money as well as religion, an expansion strategy for narrow space, a distortion of space induced by an overlap of multiple places. "Grotesque" is a result of abnormal growth, not a result of city management, and has nothing to do with urban romanticism. But it managed to change the mediocre planning and design rationality and make the city smell of deformity, morbid state, self-splitting.

Guy Debord, the French cultural critic, once professed in his best-known book—Society of the Spectacle, that the world had transformed completely from an "accumulation of commodities" to an "immense accumulation of spectacles" and that spectacles have become the essence of "media age". Where there is an independent spectacle, there will be laws rebuilt. Debord pointed out further that celebrities, recreation and tourism and urbanization had become the most prominent spectacles of this world. From my point of view, all the examples that appear in this book responded to these judgments in a dramatic way.

As an explanatory map of architecture, this book intends to lead the cultural travelers (tourists) to tour around the grotesque urban architecture and meanwhile give them a brief commentary on the functions and meaning. These comments can be bitter and direct criticism, neutral interpretation, and heartfelt praise. For a cultural traveler, this is the most interesting guide. Shanghai is a giant geographical maze and it needs an unusually valuable guide book to explain these known and unknown grotesque buildings for a justification of their existence.

The Ping'an Finance Mansion, a result of a 2.7 billion RMB investment, mixes the style of the Roman Pantheon and a modern rectangular building. Surrounded by a pile of Roman pillars and capped with a far too small Chinaman's hat-like dome at the top, it is a low-quality collage of theological spectacle and money. It is a masterpiece of Shanghai-style kitsch. And this book reveals the cultural nature of buildings of this sort.

This kind of odd collage, is a basic feature of Shanghai architecture. The "dragon pillar" at the intersection of Yan'an elevated road and North-South elevated road is an odd piece of collage made of reinforced concrete and dragon-patterned sculptural relief. The "Housing of Mountainscape" on Changde road, which applies a gigantic artificial hill on the side face of a residential building, forms a tacky and made-up collage. And the earthy yellow ancient temple – Jing'an temple casting a stark contrast to the huge body billboard next to it is a more ironic kind of collage. It is this abnormal visual combination that creates the "Disneyland"- like cartoon effect.

There is one kind collage which seems to be tolerable to us. It is a collage in the sense of time not in space. The Great World at the crossroad of Yan'an road and South Xizang road, is a collage of the old and new times. It still sits firmly in the center of a fast-changing city. Due to a jurisdiction problem, it hasn't been used for a long time. It just couldn't find an opportunity to be re-opened. Then it becomes a myth coming out of the colonial times, mixed with legends of entertainment, hooligans, actors, actresses and prostitutes. It is like a stereoscopic painting on the calendar, sitting deeply in the city center and gives us hints of the history of colonial times. As described in the book, this former indoor amusement arcade in Shanghai, featuring vaudeville, south-north traditional opera and Chinese folk art forms, is a most representative entertainment building. It will finally become a secret symbol of Shanghai landscape as it will cast its original entertainment functions to the entire city and let it smell of a teasing, naïve and clowning culture everywhere.

Another similar example is the Shanghai 1933 on Shajing road, Hongkou district. This grotesque building was once the biggest slaughter house in the Far East. Its "secretive and deep" rays shed light on complex paths. After being modernized using steel and glass, it keeps the marks of the times on the wall, like the breath uttered by those dead animals. But under its beautiful exterior decoration, its gothic gloom can't be concealed. It is a lasting smell, deeply penetrated in the bones and structures of the architecture. While not far away, a pigeon house made of bricks and cement sits on the top of an old building. The owner of the old building prepared space for each pigeon.

But the place where the pigeon house sits on is meant to be dismantled by the government so it won't be long before the pigeon house disappears. It is the common fate of the pigeon house, pigeon owner and the pigeon. But the pigeon house, together with other unapproved construction objects become the lowest, ugliest and most lively part of the city.

So far, I haven't read anything that speaks ill of such architecture. This book not only points to those grotesque land marks but also those abnormal, deformed and quickly-decaying objects, even those frowned upon by the city's managers. But it is those grotesque houses built in narrow space that screams the nature of this city. Although the city is expanding rapidly, its city center is still strong in its space tension. It forces the strange houses to grow, like a "flower of evil" that's seen as a nightmare for city administrators. The descriptions of those buildings are more like prewritten eulogies lamenting the doom of these buildings, while not so much as a humorous guide. It aims to store up the city's odd heritage in printing form.

The reason why Shanghai is called a "devilish city" by netizens is undoubtedly because of its city landscape. Architecture in Shanghai has a touch of magic that any other city in China lacks. Standing at a right spot and observe, and you will find that covered by sunshine and the fog, buildings on both sides of the Huangpu river, have a mysterious gothic image on their elevations. Pudong inherits this quality from Puxi entirely. It expands the territory of the devilish city and pushes the boundary directly to the dividing line of the land and the sea. The magical style of the "devilish city" is exactly what Shanghai's charm is about. It sends out a siren sound and seduces tirelessly the fragile souls of post-modern tourists.

Zhangjiang, Pudong
28th, November, 2013

序二：上海制造

Preface II: Made in Shanghai

塚本由晴
Yoshiharu Tsukamoto

东京工业大学教授
犬吠工作室主持建筑师
Professor, Tokyo Institute of Technology
Principal Architect, Atelier Bow-Wow

上海是世界上经历过最剧烈变动的城市之一,《上海制造》汇集了能阐明这些空间实践的特殊建筑物们。这项研究在 2012 年（《东京制造》出版后第 12 年）开始进行，令我非常高兴，此时的上海正经历着巨大的都市转型。这两项研究从单体建筑物出发，共同见证了我们的城市和社会。

我第一次来上海是在 2001 年，当时犬吠工作室应邀参加上海双年展。那时候浦东已经开始建设，但还只建成了两三栋摩天楼。在黄浦江另一侧，我们接触到了里弄中街坊们的真实生活：工人们住在他们正在建造的房屋里，很多户居民还留在面临拆迁的老房子中。这些难忘的场景让我记住了这座城市的宏大力量。2013 年是我最近一次来上海，此时相距 2010 年世博会已有三年。这座城市现已布满各种耀眼的摩天楼，变得整齐划一，但同时也趋向与其他大城市雷同。然而，无须为此惋惜，因为仍有许多见证了这座城市和这个社会变迁的建筑物存在。

上海最让我感兴趣的是，从这些建筑中能观察到多样的政治和文化背景。这正是这批建筑物的特点，同时也是“上海制造”研究的核心思想。最有名的一些案例位于前租界或外滩，它们叙述着上海在那个年代作为世界上最大国际港口的故事。甚至那些不知名的建筑物也向我们诉说着无数故事，比如共产党的诞生，新中国的住房政策，对“大跃进”的反思，有中国特色的社会主义，亚洲被卷入全球化，等等。上海在过去的若干个十年里经历了许多不同的思潮与意识形态，这一直是上海的一部分。而保留下来的这些建筑物，也都成为当中的见证者。

Made in Shanghai is the collection of the peculiar buildings which illustrate the practice of space in one of the most changing city in the world. I am very happy that this research was executed in 2012, 12 years after the publication of Made in Tokyo, in the middle of drastic urban transformation in Shanghai. These two researches share the spirit of the "witness" of the city and society through the buildings in the place.

My first visit to Shanghai was in 2001 when we, Atelier Bow-Wow were invited from Shanghai Biennale. The development of Pudong area had been already started but there were just three or four skyscrapers built. On the west side of the river, we encountered real street life of people in the neighborhood of Lilong, workers living in the skeleton of the building in which they are working, families staying in the old house under demolition. These are the unforgettable scenes which remind me the generous power of this city. My latest visit to Shanghai was in 2013, three years from World Expo 2010. The city is now made of many shiny skyscrapers and became clean and gentle in its appearance, but resemble to the other big cities consequently. But there is no need to be disappointed since there are still many buildings remain who witnessed the transformation of the city and the society.

What makes me interested in Shanghai is that the various political, cultural backgrounds are observed in the buildings. This is the peculiarity of the building in Shanghai, and could be the core concept of "Made in Shanghai". The most famous examples are the ones in the concession and the bund, who tell us the story of the world biggest international port at that time. Even anonymous buildings tell us various stories, such as about the birth of communist party, the housing policy of People's Republic of China, the reflection of iron war, capitalism in communism, globalization in Asia, and so on. Shanghai has been exposed against many different ideas, thoughts, ideologies in every few decades. The change has always been part of its nature, and the building witnesses this change if it remains.

目录 Contents

导览 Guide

上海制造：一个当代都市的类推基因
Made in Shanghai: The Analogical DNA of a Contemporary Metropolis

李翔宁
Li Xiangning

同济大学建筑与城市规划学院教授
Professor, College of Architecture and Urban Planning, Tongji University

想象城市的方法

什么是上海？或者说什么构成了一座城市？这个问题或许比我们建筑学科内长期争执不休的"什么是建筑"更加难以回答。

城市是一种复杂的综合体，城市的真实面貌或许我们永远不能真正掌握。它既是由钢筋水泥建造起来的真实的构造物，又是存在于记忆与想象中的幻象。人类如同摸象的盲人，从各自不同的体验和感知，对城市进行想象与重构，并形成对一座城市的"集体想象"。我们对于一座城市的想象，同时来自真实的体验和关于城市体验的"再现"文本。对再现城市的"文本"，如文学、绘画、电影、摄影、地图乃至城市设计中的效果图、模型，都影响着我们对一座城市意象的形成。

在传统社会中人对城市空间的体验之外，对于建构我们从未去过的地方的想象，文学和绘画起到非常重要的作用。许多经典的小说、诗歌中所描写的城市图景，已经进入人们对城市的文化想象：比如狄更斯笔下的雾都伦敦；巴尔扎克、波德莱尔笔下的巴黎；乔伊斯笔下的都柏林；乃至张爱玲、王安忆小说中的上海。同样，皮拉内西（Giovanni Battista Piranesi）的铜版画中的罗马和马奈等法国印象派画家画布上的巴黎，都成了城市景观的经典诠释。现代摄影技术将更真实的城市场景再现在人们的眼前，而电影的产生则宣告了一种全方位的关于城市的体验在电影院中呈现：在静默和漆黑的空间中，人们全神贯注于电影中所描绘的空间，而将影院外的城市，包括自己处身其中的空间抛在脑后。这些文本帮助建构了人们对于遥远地方的空间想象——我们身边的大多数人并没有去过巴黎、伦敦、纽约和柏林这样的大都市，然而这并不妨碍人们借助文学、电影、电视和各种书报杂志所汇成的宏大信息流，发展出对这些大都市极具诱惑力的想象。

人们对于城市的看法、对于城市真实的把握，在一种集体无意识的层面被自身的价值取向所重构。

价值就像一面透镜，出于阅读者自身的社会角色和生活体验，从不同的价值视角出发，变形和重塑了城市的真实。阅读者或设计城市的人，在想象城市或创造城市的整个过程中，不可能不带有自身的出发点和价值判断的影子。而城市的真实或许正是在这许许多多或接近或远离真实的阐释文本中存在，又或许这些折射和变形了的解读与想象本身就是城市的真实。在《看不见的城市》中，马可波罗向成吉思汗所描述的诸多性格各异的城市，事实上都糅合了他的家乡——威尼斯的影子。或者说，这些城市的形貌，不过是马可波罗记忆中的威尼斯一个方面的缩影，也就是他以威尼斯的基因类推出的城市。

“上海制造”

正是为了寻找这种当代上海都市基因的渴望，才有了“上海制造”这项进行中的城市研究课题。它来自我和日本建筑事务所犬吠工作室的合作。该工作室的合伙人塚本由晴、贝岛桃代和黑田润三共同出版的著作《东京制造》已经成为建筑学和城市研究的一个典范。在上海的项目中，我们试图延续这种视角和研究方法，并试图发掘出上海这个大都市独特的城市基因。这些建筑或者构筑物的基因，来自这座城市，却被加以过滤，使得我们用语汇无法描述的城市特质可以依附在它们身上得以传达。它们兼具独特性和类型性（或匿名性），用这些基因可以简单地重新组合出和现在的上海不同样貌的一座城市，但你依然能够感到这是另一个“上海”。当然，它不是17世纪时的小渔村，也不是20世纪初那个充满着怀旧情调的艺术装饰派建筑的上海滩，它是当代亚洲的一座有着独特生长模式和氛围的城市，包裹着无数或大或小或旧或新或美或丑的、无法用语言描绘的热烘烘的建筑物和构筑物的东方魔都。

我和我的助手、学生们在两年多的时间里，逐渐发掘这座城市最为人关注和最容易被忽视的都市基因：从城市地标建筑到违章搭建，从公共空间到建筑废墟，我们像旅游者一样收集城市鲜亮的名片，流连在外滩、陆家嘴和人民广场的公共空间；也像本雅明笔下的都市漫游者一样搜集着这座城市不为人所知的废弃物，出没于城市的穷街陋巷，拍摄一座座算不上建筑的临时构筑物……这些既典型又非典型的建筑和空间，彼此分离、并置、侵入、交叉、覆盖，共同组合成了一个异质混合的上海。

当然如何对我们选择的案例进行分类成为了一个头疼的问题。当下，我们选择不做实质性的分类，因为我们相信，任何不恰当的分类都将最终损害丰富性的呈现。正如博尔赫斯杜撰的貌似荒谬的所谓“中国百科全书中关于动物的分类法”：a. 属于皇帝的；b. 涂香料的……n. 远看如苍蝇的。我相信他很明白这种实际上没有分类的分类法才是对细节丰富性的最好呈现。

东方都市的当代价值

那么，我们应当如何认识上海这座城市的当代价值呢？詹明信（Fredric Jameson）在到过东亚（主要是在中国和日本的讲学与游历）之后，在文章中表达了他置身在这些城市空间和景观中所感受到的不安，尤其震惊于中国城市那种以史无前例的迅速城市化所建造的人类景观。他由此建构了未来城市景观的理论：字眼来描述——一种“肮脏的现实主义”（dirty realism）或一种“补充空间”。詹明信认为必须修改原有的“反面乌托邦”（Dystopia，描述一种政治、经济极其黑暗的假想空间，是和乌托邦的美好想象相对的），中国的城市状况是十几亿人口在经济迅速发展的同时盲目、随意地建设和消费，是一种建立在财富基础上的反面乌托邦。詹明信意识到在当今的政治、经济格局下，西方传统中产阶级主导的一种理性和有秩序的“市民社会”正面临解体和崩溃，原有的阶级、阶层的划分和身份认同被模糊化了。东方城市这种将许多毫不相关的元素和人群并置、混合在一起的史无前例的方式，或许正是对西方社会模式的危机与缺失的一种补充。詹明信认为东方城市的街道总是有点内向，因此城市作为一个整体，没有外形轮廓而成为一个巨大的、无定形的、无法描述的容器。

詹明信进而认为这种空间将补充到美国城市的空间模式中，或者说，美国城市将不得不模仿日本或中国的城市。美国电影《银翼杀手》（*Blade Runner*）中构想的洛杉矶城市的面貌，混合了高楼大厦与街边售卖汤面与日本寿司的摊点，操着混合了东西方词汇的世界语的亚裔小贩，以及街头广告的巨大屏幕上日本歌舞伎的形象，都从视觉的真实上例证了詹明信的说法。

虽然我基本赞同詹明信对东方城市的认识，但我觉得亚洲都市的类型和演化具有他尚未认识到的更积极的意义。不仅仅是对西方模式的一种补充，对当代亚洲都市的研究甚至可以在一定程度上颠覆传统意义的西方都市，而催生出一种更当代的评判城市的视角和价值观。

当然，上海和塚本所在的城市东京又有着不同的样貌：如果说东京更像是一个将不同建筑体量和空间并置后而具有不同特征的城市区域，那么在上海，这些异质的空间类型则以一种更“像素化”的方式被打散后揉在一起。在东京清晰可辨的高层、多层、低层区和大中小不同尺度的地块，在上海几乎可以在每一个地块中找到，所以上海成为了一座绵延密布着异质混杂建筑类型的都市。或者换一个角度，任选一张特定尺寸的城市街区的总平面图，根据其尺度、建筑高度和空间组合模式这些信息，你基本可以猜测出大致可能在东京哪几个地段。而这在上海几乎是不可能的，因为每一个混合了不同建筑类型的地块，彼此之间又是那么一致，几乎成为一种“通属”的地块。上海每一个地块上所容纳的基因都差不多。

“类推都市”计划

当塚本和我说起《样板房》展览的计划，并促成了我们和艺术家林明弘的合作。从某种意义上说，上海制造是对城市和居住模式的一种抽象和列举。并且我们分享着对于建筑和城市基本的价值认同：不以“美 / 丑”、“粗糙 / 精致”、“专业设计 / 自发建造”的分野等专业习惯的标准来评判，而试图直面城市的无法抽象和简化的丰富性和真实性。

最后，我们将“上海制造”项目现有的案例研究成果在本书呈中现出来。我们没有刻意修饰项目的进行时状态，相反在表现方式上以“类推都市”的方式凸显这一状态：通过抽取的基因再组合成的城市图景，经过了折射、过滤和变形，显得既熟悉又疏离，呈现为一个过滤后拼贴的异托邦（heterotopias），不同的观者都可以从中找到自己独特的上海记忆、欲望和想象。

How to Imagine a City

What is Shanghai? Or, to put it another way, what is a city composed of? This question is perhaps even more difficult to answer than that perennial topic of debate among architecture academics, "What is architecture?"

A city is a complex synthesis and we may perhaps never be able to really comprehend in full the true face of the city. It is both an actual constructed thing of steel and concrete and at the same time a phantasm that exists in memory and imagination. People are like the blind men and the elephant in the parable, imagining and reconstructing the city based on disparate individual experiences and perceptions, and from such is formed the "collective imagining" of a given city. Imagine of a city comes simultaneously from actual lived experience and from its reproduction in texts about the urban experience. "Texts" reproducing the city, be they literature, painting, film, still photography and maps, even the models and drawings of urban planners, all affect the way we form our image of a given city.

In traditional societies, besides personal experience of urban space, literature and paintings played a very important role in constructing our vision of places we had never been to. The urban landscape described in so many classic novels and poems has entered people's cultural imagination of the city: the fog-bound London of Dickens, Paris in the work of Balzac and Baudelaire, Joyce's Dublin, and Shanghai itself in the novels of Eileen Chang and Wang Anyi. Similarly, Piranesi's copper etchings of Rome or the canvases of Paris by Impressionists such as Manet have also become classic interpretations of the urban landscape. Modern photographic technology has reproduced the cityscape before our eyes in a still more realistic fashion, and the birth of film announced the presentation of a sort of full-spectrum reproduction of the urban experience in the cinema; sitting in darkness and silence, audiences are completely transported into the space depicted on the screen, putting aside any thought of the actual city outside the cinema or the space they occupy. These texts aid in the construction of our spatial imagining of distant places: most of the people around us will have not actually been to the great metropolises like Paris, London, New York or Berlin, but this does not prevent them from developing an imagining in seductive terms of such cities, and they are greatly aided by the torrent of information found in literature, film, television and a whole host of magazines.

People's opinions of the city, and their actual understanding of it, are constructed by the orientation of their values on a kind of collective unconscious level. Value is like a lens with its origins in the reader's social role and lived experience; when considered on the basis of different value perspectives, the reality of the city becomes distorted and remoulded. A reader or an urban planner, throughout the process of imagining or creating a city, will not be able to approach that unburdened by their own point of origin or uninfluenced by their value judgements. The actuality of the city perhaps exists precisely in these multiple narrative texts, be they close to or far from reality; or perhaps these refracted and distorted interpretations and imaginings are themselves the urban real. In Calvino's Invisible Cities, Marco Polo describes to Kublai Khan a varied plethora of cities, but in fact always s his tales with things drawn from his own hometown, Venice. Put another way, the descriptions of these disparate cities are really just a condensed aspect of Marco Polo's Venice of memory, cities he analogizes from the DNA of Venice.

Made in Shanghai

An urge to find the DNA of contemporary Shanghai led to the currently on-going urban research project, "Made in Shanghai". It is the product of collaboration between architectural studio Atelier Bow-Wow from Japan and me. In 1998, studio partners Moyomo Kaijima, Yoshiharu Tsukamoto and Junzo Kuroda published a similar project, Made in Tokyo, that has become a model for architectural studies and urban research. For the Shanghai project, we sought to extend the perspectives and research methodologies of its predecessor, and to uncover the unique urban DNA of the great metropolis of Shanghai. The DNA of these buildings and structures comes from this city but has been filtered, so that the special qualities of the city that we lack the vocabulary to put in words can be communicated through them. They have both their own particular nature and are of the same type (a nature that is anonymous). The DNA can be reconfigured in a simple way to produce a city that appears unlike Shanghai as it is now but that gives a sense of being an "other Shanghai". It is, of course, no longer the small fishing village of the seventeenth century, nor is it an early twentieth century Bund full of nostalgia-inducing art deco buildings; it is a contemporary Asian metropolis that has developed in its own particular fashion and has its own particular atmosphere, a magic city of the East with a clamouring myriad of buildings and structures big and small, old and new, beautiful and ugly beyond any description in mere words.

Over the course of two years, working together with my assistants and students, we uncovered both the most noticeable and the most easily overlooked urban DNA of this city: from the municipal landmarks to the illegal shanties, from public spaces to architectural ruins, we collected the bright calling-cards of the city as if we were tourists, wandering public spaces like the Bund, the Lujiazui financial district and People's Square. And like the flâneurs of Walter Benjamin's work, we also collected the detritus of a city that people do not know, photographing temporary structures hardly deserving of the name 'building' in its narrow alleys and back lanes. These standard and non-standard buildings and spaces, their separations, juxtapositions, intrusions, intersections and coverings-over come together to make the blend of different qualities that is Shanghai.

Naturally, it was quite the headache when it came to categorizing our selections. As things stand, we have elected not to create a truly substantive taxonomy, as we believe that to do so inappropriately would ultimately harm the presentation of the richness we found. It was an exercise reminiscent of Borges'fanciful "certain Chinese encyclopedia", The Celestial Emporium of Benevolent Knowledge, with its categories for animals, "a. those that belong to the Emperor; b. embalmed ones..." and so forth, until we come to "those that from a long way off look like flies". We are certain Borges too believed that only by categorizing the beasts in this non-existent typology could the richness of life be best shown.

Contemporary Values of the Asian Metropolis

How should we then understand the contemporary values of this city, Shanghai? After visiting Asia [primarily to lecture, but also to tour for pleasure, in China and Japan] Fredric Jameson wrote of the unease he felt whilst situated in those urban spaces and landscapes, in particular his astonishment at the human landscape in Chinese cities undergoing a rapid process of urbanizing transformation without historical precedent. He drew on this to construct a theory of the future urban landscape; the phrase he used being "dirty realism" or a kind of "replenished space". Jameson held that it was necessary to revise the existing term "dystopia" [which describes an imagined space where political and economic life is extremely dark, an antonym of 'utopia'] since the situation in China's cities was a dystopia built on wealth, with hundreds of millions of Chinese people living amid blind and random construction and consumption in a rapidly developing economy. Jameson came to realize that in the present political and economic order of things, the middle-class-led rational and orderly "civil society" of the

Western tradition faces disintegration and collapse, with a blurring of former classes, social strata and identities. The unprecedented way in which the cities of Asia juxtapose and mix together entirely unrelated elements and social groups are perhaps a replenishment of the deficiencies and crises of the Western model. Jameson believes that there is always a certain 'interior-ness' to the Asian city street and because of this the city as a whole lacks any outline of external appearance, becoming rather an enormous, indeterminate and impossible to describe containing vessel.

Jameson went further with this, believing that space of this kind would replenish the American urban model, or to put it another way, American cities would in future have to copy Japanese or Chinese cities. The future Los Angeles imagined in the film Blade Runner included a mix of skyscrapers with street stalls selling noodles and sushi, Asian stall holders speaking a global lingua franca that mixed Western and Eastern vocabularies, or giant screens showing advertising that employed the imagery of Kabuki performers. The visuals of this film stand testament to the correctness of Jameson's supposition.

Although I am in broad agreement with Jameson's understanding of the Asian city, I think there is a more positive significance to the types and transformations Asian metropolis that Jameson was not aware of. Not merely a supplement to the Western model, the study of the contemporary Asian city can to a certain extent overturn the traditional Western sense of the city, and give rise to still more contemporary perspectives and values to be used in urban critique.

Of course, the external appearance of Shanghai is different to that of Tokyo, the city that is home to Yoshiharu Tsukamoto. If we say that Tokyo has juxtaposed different built masses and spaces, thus creating urban districts with diverse properties, then we can find that the heterogeneous spatial types in Shanghai have been scattered and then mixed in a "pixilated" fashion. In Tokyo, one can clearly distinguish zones of tall buildings, multi-storey or low-rise constructions, or of large- or small-scale buildings; in Shanghai, examples of all these can pretty much be found in any given parcel of land, making itself a city with continuously stretching dense mixtures of buildings of all types and sizes. Considered from another angle, if you have got a general plan of a given scale at random, you are likely to figure out from which part of Tokyo it comes by making use of the information such as block scale, building height or combing mode of spaces. However in Shanghai, it is impossible in doing so, because different patches of mixed architecture are so similar to each other that they become almost "generic". In Shanghai, the "gene" of any particular block differs little from that of any other.

Analogical City

Some months ago, Yoshiharu Tsukamoto approached me to talk about the plan for the Model Home exhibition and encouraged our collaboration with the artist Michael Lin. In a certain sense, Made in Shanghai is, like Model Home, an abstraction and posited example of the city and its modes of living. I found that we at base identified with the same values concerning architecture and the city, not basing our critique on the usual specialist standards for categorization such as beautiful versus ugly, rough-and-ready versus finely-worked or professionally designed versus amateur self-building; we rather seek to face directly an urban richness and reality that is impossible to abstract or simplify.

Finally, we came to the presenting of the case studies in the Made in Shanghai research project within the exhibition space. We made no deliberate efforts to embellish the state the project was in as it was underway; quite the reverse, we emphasized this using pin-ups in the way we set out our display. At the same time as we were doing this, we also fabricated a long scroll to be displayed outdoors, Analogical City, an urban landscape created by the recombination of samples of urban DNA. Having undergone refraction, filtering and mutation, it appears at the same time familiar yet strange, presenting us with a 'heterotopia'. Different viewers can find within their own unique Shanghai memories, desires and imaginings.

重构一种异质的类推都市：“上海制造”的研究方法与视野

Reconstructing a Heterogeneous Analogical Metropolis: Approach and Perspective of "Made in Shanghai" Research Project

江嘉玮
Jiang Jiawei

李丹锋
Li Danfeng

同济大学建筑与城市规划学院
College of Architecture and Urban Planning, Tongji University

由同济大学建筑与城市规划学院师生协力完成的“上海制造”研究，在延续对亚洲大都会城市空间的总体分析外，更重要在于探究上海这座城市的内在个性。本书呈现的 54 个案例只是一个阶段成果，旨在通过这些案例展现上海尚不被了解的都市物质形态，并且激发上海民众对自己城市进一步发现、探索的兴趣。

“上海制造”研究大致分为三个主要环节。首先，案例调研基于由同济大学建筑与城市规划学院开设的“当代中国建筑与城市”国际课程，本课程指导与课的中外学生走访上海并搜集都市空间中的“异类”。让不同背景的学生共同合作的方式获得多维度的观察视角——比如，上海本地学生的视角，外地学生的视角，外国留学生的视角，等等。

发掘并搜集案例的第一环节经历了几年时间。当这部分基础调研基本结束后，由李翔宁教授领导的团队着手进行深入研究。为什么选择这些案例，这本身就已反映作为主体的调研者对上海都市空间的认知。我们在后续工作中尝试拼合投射到实物上的观念，因此尝试以制作类推都市这样的方式来重构上海。“类推上海”介乎真实与虚构之间，力图为观众呈现出一个经过折射、过滤和变形的上海意象。

本书的出版可以看作是第二环节的结束与第三环节的开始。当将这些案例推向公众面前时，我们期待公众会有热情地发掘更多的案例，不自觉地成为我们研究中广泛的参与者。这一环节试图将“上海制造”调研者的角色从高校师生转移到一般民众、从少数人转移到多数人，让更多的草根元素能够渗入已有的书卷味中。然而，我们也清晰地认识到，很多另类案例正因为它长年累月埋没于都市中才得以生存，一旦呈现于公众面前，不知道它们未来将有何遭遇。为此，我们竭力呼吁公众在发掘案例的同时积极参与到对案例的记录中来，无论通过拍摄、测量、访谈等等方式。当下的每一份记录，都将成为未来的档案。从这个角度说，出书仅仅是一个节点，而我们的研究还有更漫长的道路。

在这三大研究环节中，我们始终坚持一个明确的立场：消除惯常观点里常规的建筑与非常规的都市形态之间的差异，从而引导对所有这些都市实体进行再评估。同时，我们也致力于打破设计者与使用者之间的界限。本书所收集的这些都市碎片都是某种潜在基因，而所记录的这些使用模式均提供了一条通往理解我们的社会如何由可见之物质与非可见之关系共同形成的道路。在不以“美 / 丑”、“精致 / 粗糙”、“专业设计 / 自发建造”等惯常区分方式作为评判标准的前提下，这些案例将重新获得可资借鉴的价值，并为都市空间的创造与使用提供另外一些新思路。

在进入到具体案例之前，下面将对 54 个案例做一个简要的归类与解释，为读者归纳“上海制造”在搜集及研究案例过程中的逻辑及关系网。

归类尝试：三种维度的类别

为了帮助读者更好地理解这些案例，我们首先尝试根据某一种类别来对它们进行分类。之所以说这是一步“尝试”，是因为我们在分类过程中遇到问题——难以将某个案例放入到已有的类别中。很多案例其本身模糊的属性、暧昧的形态都导致无法按照惯常的标准来看待。因此，我们认为，显然还存在很多其他的分类方式，这里仅试图提供其中若干种维度的可能性。作为一本导览册的编制者，我们保持一种开放的态度，将更多的阐释空间留给读者。

A 类型

首先，我们采取一种类型学的方式根据各个案例自身的属性（主要是功能）来归类，区分出如下的五种类型（从 A1 到 A5）。这些类型既不独特也不新奇，不过它们会辅助我们理解一种所谓的“没有建筑师的建筑”。实际上，我们所搜集的很多案例并非简单地是一栋单体，尽管它们的外观使人产生这样的错觉。我们并不打算将这些案例称作“建筑”，而更多地采用“建筑物”或者“建造物”这样的概念来指涉它们。

A1 基础设施

基础设施被视为一座城市的骨骼或血管，它不仅扮演着传递物资、能源及信息等角色，还深刻地影响了都市人对物质实体的认知。西文“infrastructure”一词中的拉丁文前缀“infra”为“位于……下”之意，提示出该词原本指的是埋于地下的设施，大部分时候都看不见。后来，当科技发展催生出新的运输方式时，对基础设施的定义也随之扩大。机场、铁路、高速公路等成为典型的都市基础设施，与都市人的关系愈加密切。传统的建筑学研究不将基础设施列入研究名录。“上海制造”研究正好与之相反，它将都市基础设施看作最具有活力的元素，因为它通常具有很大的未利用空间。约三分之一的案例与城市基础设施直接发生关系，有些案例甚至就是某类基础设施的一部分。

A2 日常建筑

日常建筑是容纳日常生活的容器，它与都市人发生最直接的关系。一个人的社会阶层越低，他的基本生活需求通常就越高，源自本能的改造甚至是创造的欲望也越强烈。底层民众居住、饮食、休憩等日常行为给房屋带来各种更易的可能。关注日常行为与日常性、思考并回归生活的本源意义，是“上海制造”研究中的重要部分。我们的案例中有很大一部分是日常的住宅或者是住区，或者是小餐馆、菜场等每天发生日常活动的地方。也许有建筑师会质疑：这些不都是房子吗，怎么能用“日常建筑”这样的称呼？我们的立场恰恰相反，这些凝聚民间智慧的草根房屋正是当今建筑学应当包含进去的案例。通过引入“日常建筑”这个分类，本书意在消解“房屋”与“建筑”在观念上历来存在的鸿沟。

A3 标志建筑

标志建筑首先指的是一个地块里的视觉地标。与日常建筑关注细碎的场景不同，标志建筑常常与都市空间中的某种“宏大叙事”有关联。“上海制造”研究避免简单地像一般的旅游导览册那样呈现这些标志建筑，而是试图抽象出它们身上的“标志性”，也就是说，我们看重的是此类建筑在都市人的城市心理地图中所占据的分量。标志建筑被叠加其上的城市意象改造成为某种城市“代言人”。在类推上海图解的制作过程中我们发现，将遴选出来的标志建筑案例（比如东方明珠电视塔）往哪个地方一摆，那里的确趋近于上海真实的面貌。

A4 隐喻建筑

上海新近涌现的很多大型建筑似乎都趋向于寻找某种意象，而这种意象附加在建筑身上后呈现出来的是经过折射的文化性，有些时候甚至是媚俗文化。之所以分出“隐喻建筑”这一类，正是针对当下消费社会中能指泛滥的现象。图像消费造成的能指与所指的颠倒乃至错乱使隐喻建筑变得几乎与尺度无关——它只对其隐喻的内容有意义。这类建筑也构成了当下上海新奇的都市空间，折射出资本全球化过程中造就千篇一律的都市面孔，同时也看出建筑曾有的神性、宗教性、抽象性如何遭受后现代性的肢解。不过应当注意，隐喻建筑不该被当做反面例子。我们作为调研者，对消费社会有清晰的认识并保持谨慎，但本书作为导览册对所有案例并不具褒贬立场。唯有尽量保持中立，才可能激发读者继续深入挖掘的欲望。

A5 历史建筑

单独被归为一类的历史遗产或文物主要反映一种抽象的文化性。我们重点不在揭示历史建筑在创造伊始其背后的文化驱动力，而主要关注它们在空间维度上与周边都市环境保持怎样的关系、在时间维度上经历过哪些功能置换。本书中所有与历史建筑有关的案例均经历过功能上的转换，它们本身的历时性更迭不仅反映出改造者的观念，而且是上海不同时代的一个缩影。

B 操作方式 / 组合形式

在根据类型来分类的过程中，我们遇到一个棘手的问题——大部分的案例很难只放入某一类中。我们注意到它们通常是两种甚至是三种类型的组合。比如说，不管将“水闸办公室”归类为基础设施还是日常建筑都难以让人满意，因为它就像是这两种类型的混合。据此，我们试图分出一类命名为操作方式（或者是组合形式）的范畴来反映这些案例中都有哪些类型。

我们罗列了四组（每组一对）操作方式。每一组里的两种方式要么是平行的，要么是对立的。我们尝试通过这几种操作方式层析这些案例，让复合到每个案例上的各种信息逐一清晰呈现于读者面前。将遵循同一种操作方式的案例归为一类，有利于通过具体实例理解一种抽象途径。反过来看，这些操作方式也有助于解释案例的形成原因。

B1 并置

共生与杂交

对于包含两种及两种以上类型的案例，我们称这些功能共时并置于一体。共生与杂交各自描述两种不同的并置状态，它们的差别主要在于这些共存的功能相互之间在多大程度上发生关系。共生，顾名思义，意指相邻的空间或者功能之间没有很直接的联系，它们仿佛“相安无事”地共存一体。本书借用“共生”一词强调一种“平行地存在”。当数种功能构成一根链条，或者是所有权、使用权都集中在某一主体上时，我们就称呼它是“杂交”的，即形容这些功能杂合、交错在一起。总之，“杂交”一词指涉一种都市物质互相交叠的状态。静态地看，杂交或者共生的功能似乎并无大区别，但动态地看就能觉察当中差异。杂交案例中互相牵扯的空间与功能其实都有背后的利益关系，它比起内部互相独立的共生案例对外部环境有着更强的“抵抗力”。

B2 置换

植入与移除

功能置换是都市空间中最平常不过的变故，在我们的案例中，许多都经历过剧烈的置换过程。植入与移除是两种相反的置换途径，它俩描述案例所经受的某种外力作用。使用者常常会根据自身需要来植入与设计预期完全无关的功能，另外，本书收录的大量废弃案例由此前的功能被移除后造成。假如我们将案例看作一个有感知的主体，那么这种植入与移除通常都不发自其内心所愿，而为外界所迫。

B3 演化

延续与错位

跟植入与移除这两种外力作用不同的是，延续与错位这两种演化模式呈现出来的是一种源自案例自身的内力作用。通常来说，这属于案例本身自主的自我更新过程，在历史建筑这一类别中出现得很频繁。我们发现，都市街块的变更通常连带着单一案例自身的演化，可以看到，这种内力作用并不大可能只是完全的自主。

B4 包装

符号化与面具化

符号化与面具化这两种包装手段是专门针对消费社会出现的“新”、“奇”建筑而提出来的一组操作方式，它们主要描述的是这层“消费外衣”如何被附着到都市案例之上。标志建筑与隐喻建筑之所以往往会成为都市中的某种符号或面具，包装这一种手段起到很重要的作用。

C 规模

极小—小—中—大—极大

区分出类型与操作方式这两种维度后，我们再根据规模来对全部案例进行排序。在此我们共分五级，从极小经由小、中、大一直到极大。本书中，从类似鸽舍这样的迷你宠物建筑尺度，一直到类似虹桥枢纽这样的超大型都市枢纽，各种规模的案例都有，尺度上的差异造就案例生成结果的多种可能。日常建筑的规模往往都不大，而基础设施或标志建筑通常有着较为庞大的规模。当不同的体量规模确定后，功能规模也基本能确定。

自上而下抑或自下而上：案例如何形成

如上我们提供了三种维度的分类可能，这种分类非常有助于理解这些另类案例如何形成。我们以下通过列举若干个案例来阐释这些异质空间如何被创造出来。我们区分出两种生成案例的途径：自上而下的“设计”与自下而上的“非设计”。“上海制造”调研探究这两种途径各自会如何造就一个案例，以及这两种途径在多大程度上存在融合的可能。

从数量上看，日常建筑是我们案例中的大类，它最常见的操作方式就是共生与杂交。日常建筑中这种自发的改造行为一般都是自下而上的，远离任何官方的力量。在都市大环境中，基础设施通常都有巨大且未被利用的体量（例如案例 14、16 与 50 等），这使得其他类型特别是日常建筑很有可能添加进来。高架路下的综合体是一个系列的研究，在具体的案例介绍中能看到，两个高架综合体案例（案例 12 与 32）分别可以作为共生与杂交两种操作方式的典型。

我们还能看到一些交叠了几种功能但错时活跃的案例，比如三明治学校与菜场旅馆（案例 25 与 40）。一种功能异常活跃时，另一种功能往往在睡眠。可以说，在时间段上错位的需求促使这几种功能自主地组合到一起，互相磨合与适应。这种源自基本功能需求所带来的结果也反映出自下而上的更多可能性。

标志建筑与隐喻建筑通常都是自上而下设计出来的结果，有些时候甚至是非常“官方”的设计。它们一般来说不大会与其他类型结合，呈现出某种与四周隔绝的状态（例如案例 01、04 与 06）。显然，“官方”的力量通常都不大愿意去打破设计时的既有界限。此外，资本在自上而下的过程中往往扮演着重要角色。能够看到，一旦资本撤出，某些案例即刻陷入衰颓的废弃状态（例如案例 27 与 39）。

自上而下与自下而上在某些案例上存在融合的可能，高架综合体 I 就是这两股力量交汇的结果。这个高架路底下的综合空间并非只是底层市民的自发行为，还包括当地街道办事处的大力组织。当年建造菜市场时街道办事处甚至还请过设计单位绘制图纸。街道办事处作为政府基层部门介入到这类空间的改造与利用中，说明自上而下的“设计”与自下而上的“非设计”可以找到某一个契合点。

我们赞赏并尝试记录来自底层、来自大众的力量，同时也关注权力影响下的空间设计及功能规划。虽说“上海制造”研究并非主要在解释案例的成因，但我们依然希望读者能从中获得追根溯源的兴趣。

重新评估当下亚洲大都市中的异质空间

从 2012 年 3 月上海外滩美术馆举办的“样板屋”展览开始，“上海制造”研究开始逐渐走向公众。我们在展览过程中听到很多反馈声音，也越来越意识到，上海之外的许多其他亚洲大都市同样拥有相似的另类都市空间，它们也都能成为上海的一面镜子。在 2013 年深港“城市 / 建筑”双城双年展上，三个东亚大都会——东京、上海、香港——的都市异类空间研究项目并排展示。作为本次双年展“城市边缘”主题的一部分，这三个展位所关注的相似都市空间、所采用的相似表现形式（例如轴测线条图）反映出这三个城市的研究团队有着相近的方法与视野。在此，我们认为，很有必要对当下亚洲大都市中的异质空间作出重新评估。

亚洲大都会的都市空间调研应当做成一个系列。从东京到上海，到香港、新加坡、孟买乃至更多的亚洲都会，高密度的都市空间构建起高密度的都市生活。相比起传统的欧洲城镇，表面上秩序混乱的亚洲都会其实内含各种民间的逻辑与智慧，创造出另类的都市空间与功能。以“上海制造”为例，很多案例的未来都处于一种非常不确定的状态。之所以不确定，是因为在当下中国的城市法规中这些案例都很难有一个明确的身份。以左翼知识分子的观点来看，记录并研究这种草根案例或许也可当作是一种对平权的呼吁，打破原有划分社会阶层的界限，而重新评估这些案例表达了对抗衡极权的诉求。此外，当对比这几个城市的土地所有制以及政府的土地政策时，我们能更深入地理解这些都市异质空间。从东京到上海，从土地私有到土地国有，许多案例均由土地的细分造成。反过来，整合土地的方式也有很多种。在上海，将零碎地块合并起来的手段要比东京猛烈很多，这反映出土地国有制具有强大的土地资源整合力，但对底层民众自发搭建的建造物来说却是一种灾难。“上海制造”研究指向更为深入的社会架构研究。

发现并理解一座城市是一条漫漫长路。我们只是尝试记录当下各种非常规的功能与空间，重构一种异质的类推都市。希望本书在此为各位读者打开一扇通往异质上海的想象之门。

"Made in Shanghai" project is done by a research team consisting of teachers and students from Tongji University's College of Architecture and Urban Planning ("CAUP" for short). Along with being a continuous analysis of the urban space in Asian metropolises, this research mainly focuses on the internal characters of Shanghai. The 54 cases are presented throughout this book. Through these cases, the project aims at discovering the urban physical forms that are still unknown to the public, while also trying to interest the public in exploring their own city.

"Made in Shanghai" research, in general, can be divided into three phases. In the very beginning, the survey and collection of these cases is based on an international course given by CAUP in Tongji University, which is titled "Contemporary Architecture and Urbanism in China". It instructs and encourages the participating Chinese and foreign students to walk through Shanghai and gather these diverse cases. A cooperation of work, between both Chinese and foreign students from various backgrounds, aims at obtaining multi-dimensional perspectives. These perspectives include that of native Shanghainese students, of students from other Chinese provinces, and of students coming from abroad.

It has taken several years to discover and collect the cases for the first phase. When the basic survey came to an end, the research team led by Professor Xiangning Li took on a further study. The next phase of cases continues with the reflection of the thoughts from the researchers on Shanghai. In this phase, we try to combine the concepts projected on the cases, in turn, reconstructing Shanghai by creating a diverse analogical metropolis. "Analogical Shanghai" is between reality and fiction; it presents the audience with an image of this modern city through refraction, filtering, and deformation.

The publication of this book can be regarded as the end of the first two phases and the start of the third one. When all these cases are presented to the public, the public will become part of the research team by helping in the enthusiastic discovery of more cases. The switch of the researchers' role from college teachers and students to the general public, from the minority to the majority, will bring more local elements into the existing "scholarliness". However, it is clear that "Made in Shanghai" may probably pose a "risk". Many of the cases exposed in this book can only exist when submerged into the tremendous urban space because, according to the current legislation of urban planning, they are regarded unapproved. Once they are presented in front of the public, the futures of these projects are unknown. It is so common to dismantle a building in the current rapid urban development; therefore, we call on the public not only to discover the cases but also to take an active part in recording them. These cases can be recorded through photographing, measuring, or interviewing; each little bit today will become a file for the future. It is our hope that the publication of this book could be a milestone, but our research is far from the end.

In these three main phases, we hold a clear stance of eliminating the distinction of normal architecture and abnormal urban forms from the conventional viewpoint, thus leading to the reassessment of all these physical urban entities. Also, we devote ourselves to break the border between a designer and a user. All these urban fragments we have collected are potential genetic factors of this metropolis, in wihich all the utilization modes of the cases we have recorded provide a pathway to understand how our society is formed both by visible substances and invisible relationships. Our critiques are not based on the customary standards for categorization such as beautiful versus ugly, rough-and-ready versus finely-worked, or professionally designed versus amateur self-building. We consider it possible to gain new values from these cases and provide other new ideas for the creation and utilization of urban spaces.

This article will give a brief explanation of the categories of the 54 cases before going into each one individually. In this way, the logic and network of the process of collecting and studying cases of "Made in Shanghai" is shown.

An Attempt to Classify: Three Aspects

In order to help our readers better understand the cases, we attempt to classify them in certain categories. We regard it as an "attempt" because we encounter a problem when classifying, that is, we find it difficult to group some of the cases. It is hardly possible to judge the vague properties and ambiguous forms of many cases by common criteria. So, we agree that there still exist many other sorts of categories, and this article merely attempts to provide a possibility of some dimensions. As the maker of a guidebook, we keep an open attitude and hope to leave space for interpretation for our readers.

A Types

At first, we adopted a method of typology to classify all these cases according to its own property, mainly their functions. In the following from A1 to A5, five types have been made. None of these types are unique or novel, however, they help understand the so-called "architecture without architect". In fact, most of the cases collected are not a single building even though they appear as one from the outside. Thus, these cases shall be called "buildings" or "structures" rather than "architecture".

A1 Infrastructure

Being regarded as the skeleton or vessels of a city, physical infrastructure has not only played the role of transmitting goods, energy, or information, but also deeply influenced urbanites' conception towards physical entity. The Latin prefix "infra", which means "below", suggests that the term originally referred to those underground facilities, most of the time invisible to the public. Later, when new transportation modes appeared due to the development of science and technology, the definition of infrastructure expanded. Now, airports, railways, and high roads are typical examples of urban infrastructure which has a closer relation to urbanites. From the viewpoint of traditional architectural studies, infrastructure is not included in the list. On the contrary, "Made in Shanghai" research treats urban infrastructure as the most dynamic element, for it usually has large unused space. About one third of the cases have a direct relation with urban infrastructure, and some are even a part of it.

A2 Everyday Architecture

Everyday architecture is a container of everyday life and has the most direct relation to urbanites. The lower class of society a man is in, the higher his basic life demands usually grows. He usually has a greater desire, which is derived from his instinct to make alterations or even creations. The everyday behaviors such as living, dining and entertaining have brought all kinds of possible alterations to the building. "Made in Shanghai" research is concerned with everyday behaviors, as well as, thinking about and returning to the essential meaning of living. A large part of our cases is ordinary house, residence, or places such as small restaurants and food markets, where everyday activities take place. Some architects may doubt, even question: Aren't they buildings? How can you call them "everyday architecture"? This book holds an opposite stance, in which such houses of grass roots are a cohesion of folk wisdom. In fact, they are exactly the cases that should be included in current architecture. By introducing the type of "everyday architecture", this book aims at erasing the longtime existing gap between "building" and "architecture".

A3 Landmark Architecture

Landmark architecture refers to the visual landmark within a block. Different from the divided and fragmented scenes in everyday architecture, landmark architecture usually links with the "grand narrative" of urban space. "Made in Shanghai" research tries to avoid presenting them simply as a common tourists' guidebook, but attempts to extract "the symbolic" from them. In other words, this book are concerned with the place such architecture takes in the mental city map of urbanites. The city image overlaid onto landmark architecture modified it into a city's "spokesman". In the process of making the diagram of Analogical Shanghai, researchers found that where the landmark architecture was placed (such as the Oriental Pearl TV Tower) created a map that seems like the real Shanghai.

A4 Allegorical Architecture

It seems that many of the newly completed large architecture in Shanghai tend to seek a certain image, which reflects a kind of refracted culture and sometimes even a kitsch one. It is the signifier-charged excesses of today's consumer society that made the type "allegorical architecture" distinguish. The inversion and even disorder of signifier and signified by the consuming of images has made this type of architecture hardly have anything to do with scales, in which it is merely meaningful to what it symbolizes. "Allegorical architecture" also constitutes the current novel urban spaces in Shanghai, refracting the repetitive urban faces made by the globalization of capitals. At the same time, one can find how the deity, religion, and abstraction of architecture are being dismembered by post-modernity. Nevertheless, allegorical architecture should not be treated as a contrary example. As a researcher one keeps a clear and cautious stance towards consumer society, however, this guidebook does not have any judgment upon all these cases. Only by remaining neutral can this book arouse the desire of the readers to further their discovery of the cases.

A5 Historical Architecture

Historic heritages or relics mainly reflect a kind of abstract cultural issue. The focus of this book is not to reveal the cultural driving forces behind such historical architecture, but to elaborate both the relations of these cases with surrounding urban environment and the change of functions. All the cases related to historical architecture in this book have experienced functional modifications. Their changes not only reflect the modifiers' conception, but also become the essence of different eras in Shanghai.

B Operational Modes/Combining Forms

In the process of classifying the cases according to their types, a dilemma was encountered, in which most of the cases could not be placed into a certain group. It was noticed that these cases are often a combination of two or even three types. For example, the case "Sluice Office" could be classified as either infrastructure or as everyday architecture, however, for it seems like a combination of these two types. Under the circumstance, another category was developed. This category is named operational modes (or can be called combining forms) to reflect how many types and how various types exist in a single case. Four pairs of terms are listed as the operational modes. The two terms within each pair are parallel or opposite. The researchers tried to extract all kinds of information that has been overlapped onto each case by adopting these operational modes. When grouping the cases according to their operational modes, one gains an understanding of these abstract approaches through vivid examples. Conversely, such modes can also help explain how these cases come into a being.

B1 Juxtaposition
Coexistence and Hybridization

When a certain case contains two or more types, these functions are juxtaposed. Coexistence and hybridization describe two different ways of juxtaposition, in which they are mainly differentiated by the extent to which the functions are related to each other. Coexistence, just as its name implies, indicates that the nearby spaces or functions have no direct relation to each other and are in harmony. The word "coexist" to used to emphasize a condition of "living in parallel". When several functions come to form a chain or the ownership and the right to use are drawn into a common subject, it is called "being hybridized", which describes a condition of being mixed and intertwined. In other words, "hybridization" refers to a state of overlapping of urban entities. The intertwined spatial or interest relationships within hybridized cases usually bring stronger resistance than the coexisting cases.

B2 Replacement
Implantation and Removal

Change of functions is one of the most common events in urban space. Many of the cases have experienced extreme functional changes. Implantation and removal are two contrary ways of replacement, in which the effect of a certain external force applies the changes. The users often implant a function which is totally beyond the expectations of the original design. Moreover, many of the cases in this book have already become abandoned, which is caused by the removal of the former functions. A case is often treated as a subject with perception, most of the time these subjects are not willing, but have to accept such replacements that were imposed on them.

B3 Evolution
Extension and Misplacement

Different from implantation and removal, extension and misplacement describe an internal force derived from itself. Generally speaking, it belongs to the automatic self-renovation of the cases themselves. Such evolution often takes place in historical architecture. It was discovered that, the changing of an urban block is usually accompanied with the evolution of a single case. From this, it can be assumed that such effect of internal force would not be totally autonomous.

B4 Dressing
Signifying and Mask

There are two ways of dressing: one, to signify and two, to mask. These are a kind of operational mode especially for the "novel" architecture in consumer society. They mainly describe how this "overcoat of consumption" is casted onto the cases. The mode of dressing plays an important role in the process in which Landmark and Allegorical Architecture are becoming an urban sign or mask.

C Scales
Micro - Small - Mediate - Large - Macro

After having differentiated various types of operational modes, the researchers started to rank all the cases according to their scales. They were divided into five grades: from micro, small, mediate, large, and to macro. This book includes cases ranging from the pigeon house, which is a kind of mini pet architecture, to the super scale of Hongqiao Complex. The differences in the scales of cases result in the various possibilities of how they come into being. The scale of everyday architecture is often small, while that of infrastructure or landmark architecture is usually very large. When different scales of the volumes have been made clear, the scales of functions can be determined.

Top Down or Bottom Up: How these cases come into being

As shown above, the researchers provided a possible form of classifying in three dimensions. This helps understand how the special cases come into being. In the following, several cases were taken as examples to illustrate how these various spaces are created. There are two modes: the "design" from top down and the "non-design"from bottom up. "Made in Shanghai" explores how these two modes create a case, and to what extent are the two modes merged together.

Everyday architecture has the largest amount of cases, and the most common operational modes are coexistence and hybridization. The most initiative transform in everyday architecture is often from bottom up, which is frequently far away from any official forces. Most of the time, urban infrastructure creates gigantic and unused volumes, which makes it possible for other types, especially everyday architecture, to fill in (for example case 14, 16 and 50). The complexes beneath the elevated roads can be made into a series of research. Case 12 and 32 can be treated as the typical case of coexistence and hybridization respectively.

There are also some cases that overlap with several peak-shifting functions, for example the Sandwich School and Market Hotel (case 25 and 40). A function is usually at sleep when another is dynamic. The demand of peak shifting functions over time has prompted several programs to combine and adapt to each other. The result of this basic demand for functions reflects the possibilities of bottom up.

Landmark and Allegorical Architecture are the outcome of designing from top down. Generally speaking, it is less probable that these types will combine with other types, thus they are often in an isolated situation with its surroundings (such as case 01, 04 and 06). It is obvious that the so-called "official" forces are usually not willing to break the boundary of design. In addition, capitals have played a very important role in this top-down process. It can be found that once the capitals have moved out, some cases will immediately fall into a situation of abandonment and depression (case 27 and 39).

In some cases, there exists possibility of the mergence of the two modes – top down and bottom up. The complex under the elevated road Ⅰ is one case showing the joining of these two forces. This complex wasn't merely the result of the spontaneous behavior of local citizens, but due to the effort by the local street office. They even asked a certain design studio to make the drawings of the market. The role played by local street offices as a grass-root government department into the renovation and utilization of such spaces indicates that the top-down "design" and the bottom-up "non-design" can reach a certain integrating point.

The researchers not only admire, but try to record the strength ascending from bottom up, which comes from the people. It is also important to pay much attention to the spatial design and functional planning under the influence of power. Throughout "Made in Shanghai", the researchers did not focus on explaining cases, but on making the readers interested in tracing the source.

Reassessment of these Heterogeneous Spaces in Current Asian Metropolises

Since the Exhibition "Model Home" in Shanghai Rockbund Art Museum in March 2012, "Made in Shanghai" research began going public. The researchers have received a lot of responses from the public, and it has become aware that, the other Asian metropolises also have similar assortment of spaces of their own, creating a mirror for Shanghai. In the 2013 Shenzhen-Hong Kong Biennale of Urbanism and Architecture, research projects of three East Asian metropolises, Tokyo, Shanghai, and Hong Kong, were displayed next to each other. As part of the biennale titled "Urban Border", these three cities' stands all focus on similar urban space and adopt similar representation method. For example, the axonometric drawings reflect the similar approach and research perspectives of the three cities. Therefore, it is important to reassess the various spaces in the current Asian metropolises.

The survey on space based on Asian metropolises could be made into a series. From Tokyo to Shanghai, to Hong Kong, Singapore, Bombay and other metropolises, the congested lifestyle is being constructed by high-density urban space. Compared with traditional European cities, these Asian Metropolises actually contain folk wisdom, thus creating abnormal urban spaces and functions. Many cases of "Made in Shanghai" have a very uncertain future as a survivor in this rapid-developing urban environment. From the view of the left wing, it is an appeal to equality to record and study these grass-root cases, longing to break the borders of social classes. It reflects a plead against totalism to reassess all these heterogeneous cases.

Moreover, when comparing the ownership of the land in these three metropolises, one can reach a further understanding of the heterogeneous urban spaces. From Tokyo to Shanghai, from private land ownership to land nationalization, many cases are caused by the subdivision of land. On the contrary, there are many other ways to integrate land. In Shanghai, it is more fast and aggressive to integrate the land than in Tokyo; which in turn, reflects the strong state ownership of land and how it has a strong integration of land resources. However, it is a tragedy for the self-building of grass-root inhabitants. "Made in Shanghai" research points at the further study of social frames.

It is an endless road towards discovering and understanding our own city. We are merely trying to record all the current abnormal functions and spaces, and to reconstruct an analogical metropolis. This guidebook opens a door of imagination for its readers to a heterogeneous Shanghai.

导览
Guide

场所：浦东新区世纪大道 1 号
功能：电波发送 + 观光 + 博览 + 会议
Site: No.1 Century Avenue, Pudong New Area
Function: broadcasting + sightseeing + exhibition + conference

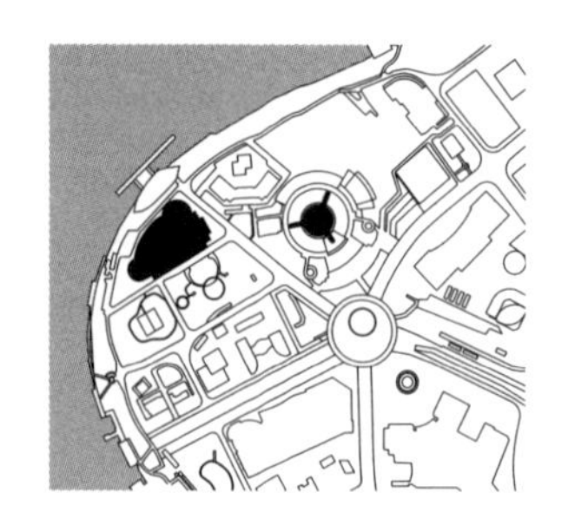

01 东方明珠广播电视塔 The Oriental Pearl TV Tower

东方明珠广播电视塔坐落于黄浦江边，浦东新区的陆家嘴之尖，与外滩万国建筑群隔江相望。它总高 468 米，主体结构高 350 米。上海城市历史发展陈列馆位于地下一层的零米大厅，90 米高的下球体为太空游乐场，263 米高的中球体为主观光层，350 米高的上球体为太空舱。电视塔旁边是上海国际会议中心，专门举办大型国际会议、商务论坛。

The Oriental Pearl TV Tower is located on Huangpu River and at the tip of Lujiazui, meeting the exotic building clusters in the Bund face to face. The tower has a total height of 468 meters, of which 350 meters belongs to the main structure. In the grand hall on the zero-meter level lays Shanghai Municipal History Museum. The main sightseeing floor is situated on the 263-meter level with a capsule being on the 350-meter level. Beside the tower stands Shanghai International Conference Center, specializing in international conventions and business meetings.

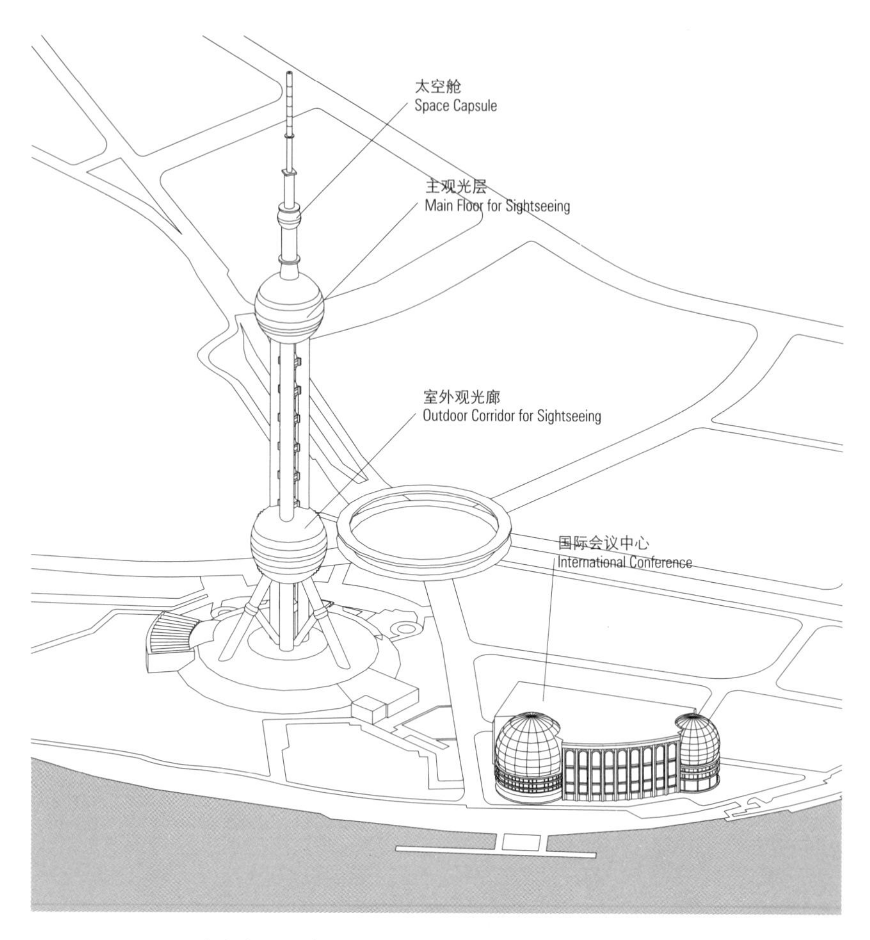

在任何一张有关浦东摩天大楼群的图片里，总能找到东方明珠广播电视塔的身影。在图像消费的时代里，建筑物能够被海量的图像信息包装成一个城市乃至国家的“代言人”。作为上海的标志性建筑，东方明珠依凭雄壮的体量以及“大珠小珠落玉盘”的外部造型寓意，攫取了外滩一带的视觉统治地位。它成为大众媒体宣传洪流中的“媒介发射器”。

The Oriental Pearl TV Tower can be found in any current picture of the Pudong skyline. In this image-consuming society, architecture can be packaged by numerous images as a "spokesman" of a city and even of a nation. Being an iconic building in Shanghai, the Oriental Pearl TV Tower gains the visual dominant position by its gigantic volume and allegorical external shape of "various-sized beads falling onto a jade plate" . It has become iconic in the propagandistic floods of public media.

场所：静安区延安中路 1000 号
功能：展览 + 会议
Site: No.1000 Middle Yan'an Road, Jing'an District
Function: exhibition + conference

02 上海展览中心 Shanghai Exhibition Center

上海展览中心原名“中苏友好大厦”，为纪念中苏友谊建于1955 年。它是中苏友好时期，上海在苏联专家的帮助下建造的第一座展览建筑，组织和举办过数百个国内外展览会，接待过许多外国元首和政府首脑，是上海乃至中国重要的对外交流窗口之一。它坐北朝南，主楼矗立正中，与两侧辅楼围合出前广场，整体设计具有俄罗斯古典主义建筑风格。

The former name of Shanghai Exhibition Center was "Sino-Soviet Friendship Mansion" . Erected in 1955 in memory of the friendship of PRC and USSR, it was the first exhibition architecture in Shanghai under the assistance of the experts from USSR. Hundreds of domestic and international exhibitions have been held here and it has also received many chiefs of foreign governments, thus making itself one of the most important windows of external exchange for China. It is located facing south and backing north. The front square is formed with the main building situated in the middle and two wings on both sides. A Russian classical architectural style was adopted in its design.

热烈庆祝中华人民共和国成立64周年!
KONKA
观众票务中心
Ticket Office
票价10元
①
②
③
④

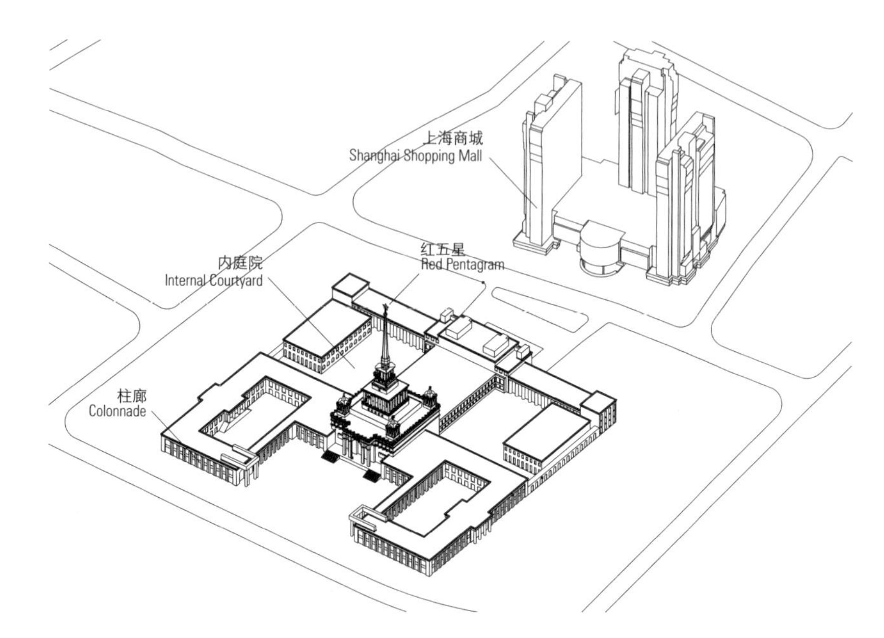

历史建筑具有触发内心怀旧情绪的特征。红色年代、中苏交好的印记深深烙在上海展览中心这栋宏伟的建筑上，触动经历那个年代的人的记忆。然而在年轻人心中，它可能只是一座彰显异国情调的建筑。时间会瓦解一栋建筑历史上曾有的政治内涵，而上海当下的时髦社会则将它无情地卷入对风格无止境的消费中去。

Historical buildings are characterized by stirring up a reminiscent mood. The impression of Sino-Soviet friendship is deeply imprinted on Shanghai Exhibition Center, arousing the memory for those who once experienced that period. However, it may seem to be merely a building manifesting exotic styles in the eyes of the youth. The politics that once was within the history of a building is now being dismantled by time. Shanghai, this modern society, is ruthlessly drawing this building into the endless consuming of styles.

友 谊

场所：静安区南京西路 1376 号
功能：办公 + 商业 + 休闲娱乐
Site: No.1376 West Nanjing Road, Jing'an District
Function: office + commerce + leisure and entertainment

03 上海商城 Shanghai Center

1990 年建成的上海商城位于上海展览中心背后，由美国建筑师约翰·波特曼设计。整座建筑呈“山”字形，由一个结合中庭和沿街退台形式的裙房将三栋主楼连接起来。入口广场上设有宽阔的前庭院，由露天楼梯引人流上楼，所见之处通透明亮。庭院四周设置了一组组商铺及餐厅，从庭院通过自动扶梯或楼梯可直达宽敞的大厅。楼顶有花园，还有直升机停机坪。

Shanghai Center lies behind Shanghai Exhibition Center. It was designed by the American architect John Portman and completed in 1900. The whole architecture has a layout similar to the letter "E", in which three main high-rise buildings are connected by an outskirt one combining the central courtyard and the set-backs facing the street. A broad front courtyard is located in the square at the entrance, and the grand outdoor staircase leads streams of people up to the first floor, where is bright and transparent. Groups of shops and restaurants are placed around the courtyard, from where one can arrive at the main hall directly by escalators or stairs. There are gardens and a helipad on the roof of this grand building.

ROLEX

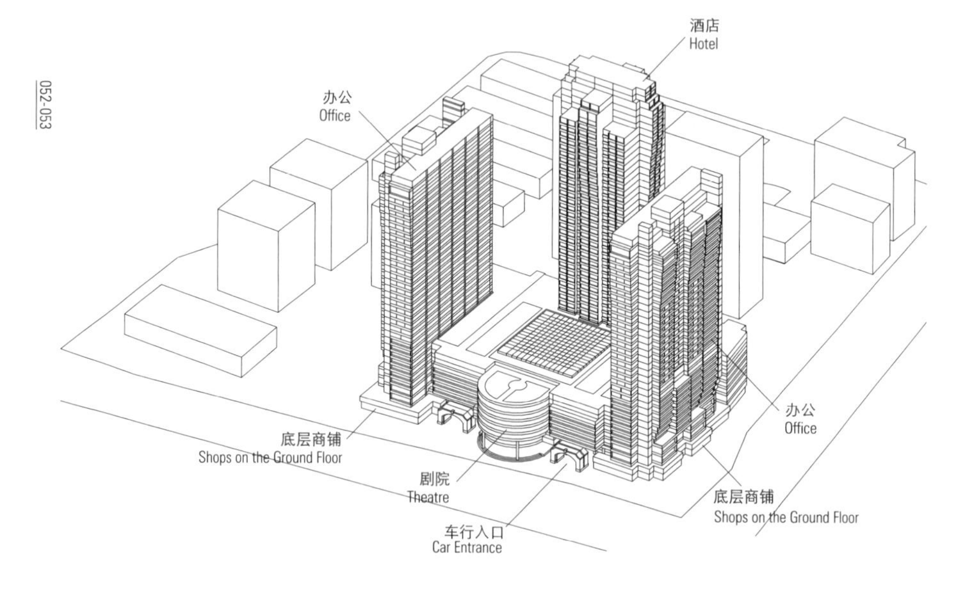

当代商业建筑依凭资本运作已俨然形成充斥于日常生活中的一道“奇观”。商业消费首当其冲，而美国都市中不断演化的商业空间更深刻地改变着人群的日常消费行为模式。波特曼作为曾经最成功的美国商业建筑设计师，通过上海商城这个项目首度将资本主义世界的商城空间模式引入中国，促进资本奇观社会的全球化。

The current capital-driving society has already become a marvelous "spectacle" in our everyday life. Commercial mode is the first to be affected, and the developing commercial spaces are deeply changing the consuming behavior in the metropolises in the USA. Portman, as once the most successful American architect of designing commercial spaces, brought such spatial mode of commerce into China through this project, promoting the globalization of spectacular capital-driven.

SHANGHAI
ROLEX
MASTERS
上海劳力士大师赛

场所：浦东新区陆家嘴环路 1333 号
功能：办公
Site: No.1333 Lujiazui Ring Road, Pudong New Area
Function: office

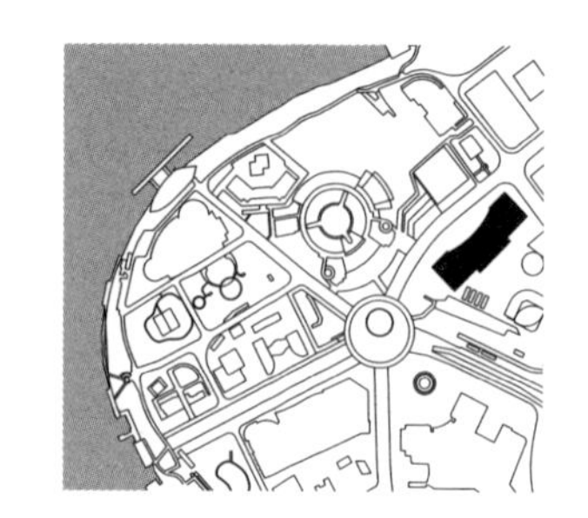

04 平安金融大厦
Ping'an Finance Mansion

平安金融大厦坐落于浦东东方明珠广播电视塔旁，地下 3 层，地上 38 层，高 198 米，总建筑面积 16.3 万平方米。它包括各类办公室与会议室，并配有接待大厅与三层地下停车场。一个巨大的穹顶覆盖在退台状的建筑主体之上，立面上排布得密密麻麻的装饰柱子使它获得“千柱楼”之称。

Ping' an Finance Mansion is located beside the Oriental Pearl TV Tower in Pudong. It has 3 underground floors and 38 aboveground floors, with a height of 198 meters and a total floor area of 163,000 square meters. It contains various offices and conference rooms, and a reception hall as well as a three-story underground parking lot. A huge dome is placed on top of the set-back of the building, and it was called "a building of thousands of columns" because of the thickly dotted decorative columns on the façade.

星展银行 DBS

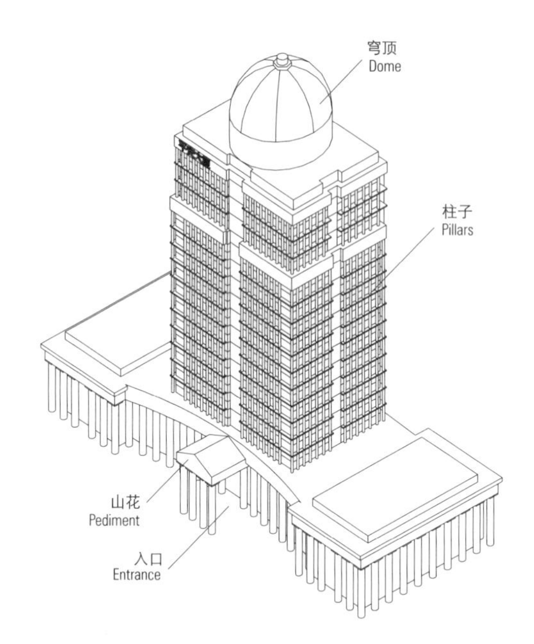

平安金融大厦是当代上海社会对风格消费的又一典型实例。立面上不承担结构作用的仿爱奥尼柱式以及每层柱子之间的檐口线脚，无一不标示这栋建筑力图打造的欧陆古典风情。裙房高达 20 米的仿多立克柱式构建出一种象征空间，以震撼眼球的媚俗实体瓦解了它模仿的原型——古希腊或古罗马神庙——所具有的神性。通过对古典形式与风格的挪用与再造，平安金融大厦为它的业主重构了一个隐喻资本、权力与社会高等阶层的空间体。

Ping'an Finance Mansion is another typical case of the consuming of styles in contemporary Shanghai society. The columns are of the pseudo-Ionic Orders (they do not bear the main structure). In between the columns are horizontal cornices and architraves. The architectural details are derived from the European classical style, which the Finance Mansion is trying to achieve. The 20-meter high pseudo-Doric Orders on the ground floors of the skirt building construct a kind of symbolic space, leading to the disintegration of the deity of the prototypes it imitates—ancient Greek or Roman Temples—through an entity of babbittry and kitsch. By taking classical forms and styles but recreating them inferiorly, Ping'an Finance Mansion reconstructs a kind of space of metaphors referring to capital, power and high social echelon.

汇亚大厦
中国平安

场所：普陀区常德路 1258 号
功能：居住
Site: No.1258 Changde Road, Putuo District
Function: residence

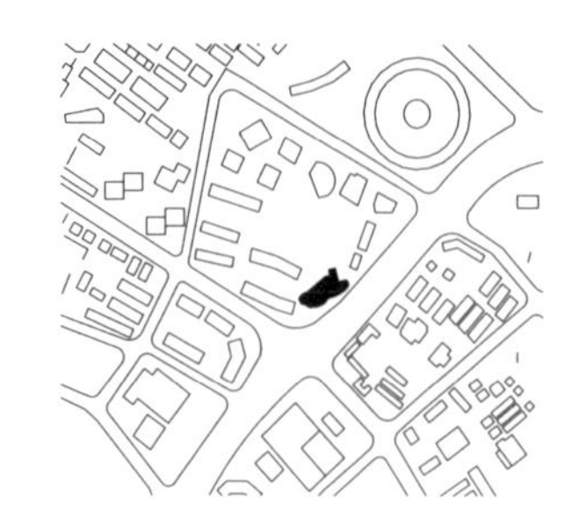

05 山景房 Housing of Mountainscape

上青佳园是上海的中档住宅小区，2001 年竣工，能供 1 800 户居民入住。开发商为满足售楼时提出的小区景观“有山有水”的承诺，在一期完成后，改造了沿长寿路的公寓楼外立面，将其南翼用混凝土假山覆盖，内部还藏有电梯。假山直接面向长寿路主干道，人们无论在小区的内部还在外部都能一眼看到它。

Shangqing Residential Area is an intermediate-level housing property in Shanghai, completed in 2001 providing residence for 1,800 families. In order to comply with the promise of mountainous and aquatic landscape when selling the apartments, the developer renovated the façade along Changshou Road and covered the south wing with an artificial concrete hill. Inside the hill there is an elevator. The concrete hill faces towards Changshou Road directly and can be caught sight of by anybody no matter inside or outside this residential area.

上海城建 隧道股份
5

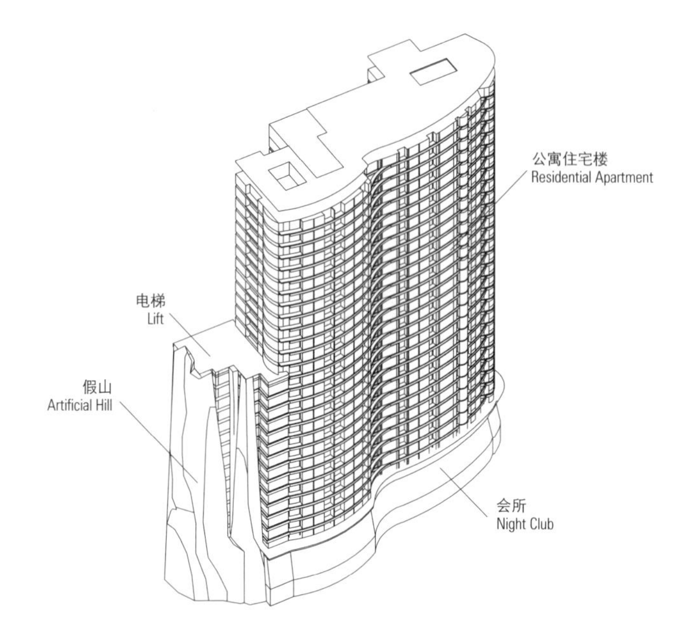

“山景房”名字的来由就是这尺度巨大但不宜人的假山。它的外观与内部功能严重脱离，开发商此举最终沦为一种虚假的装饰行为。从“有山有水”的口头或文字许诺到最终的大型假山实体落成，资本纵容着从“能指”到“所指”之间荒诞的错位。在消费社会中，“能指”可以被用作销售时的噱头，描绘一种田园风光；而“所指”可以偏离它所该到达的“能指”，并制造出一种媚俗实体。

The name "the Housing of Mountainscape" comes from this large-volume but unfavorable artificial hill. The severe separation between its external shape and internal function indicates the fictitious decorative practice by the developer. From the oral or written promise of "mountainous and aquatic landscape" to the final gigantic artificial hill, capital produces the ridiculous misplacement of "the signifier" and "the signified". In consumer society, "the signifier" can be used as a selling point to describe a kind of Arcadia; meanwhile "the signified" can deviate from its supposed target, thus making a kind of kitsch.

场所：徐汇区虹桥路 100 号
功能：住宿 + 休闲娱乐
Site: No.100 Hongqiao Road, Xuhui District
Function: accommodation + leisure and entertainment

06 西藏大厦万怡酒店 Tibet Mansion of Marriott

西藏大厦万怡酒店是美国万豪集团旗下的第 800 家万怡酒店。它开业于 2009 年，共 23 层，含客房 364 间。酒店配备有各类提供客人使用的休闲娱乐设施，并毗邻徐家汇天主堂与旧法租界区。为呼应藏式建筑的风格，该酒店在立面上使用藏式花窗，并且将一个藏式宫殿建筑的传统大屋顶覆于顶上。

Tibet Mansion of Marriott in Shanghai is the 800th hotel of Marriott Group from the USA. It started business in 2009, and has 23 floors containing 364 guest rooms. The hotel provides guests with various facilities of leisure and entertainment. Also, it is adjacent to Xujiahui Catholic Cathedral and the former French Concession. In order to reflect a kind of Tibetan architectural style, the hotel adopts Tibetan lattice windows on the façade and places a traditional big roof of Tibetan palace architecture on top of itself.

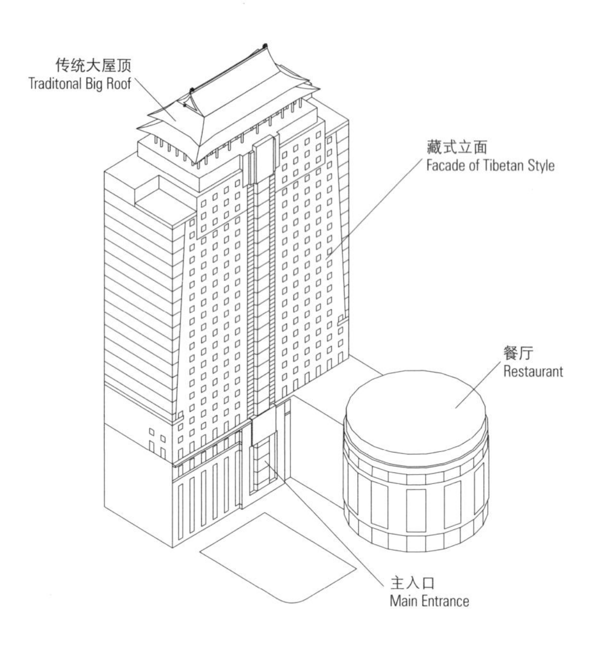

这栋头戴大帽，身披盔甲的西藏大厦仿佛将作为一个群体的布达拉宫压缩成一个单体。与平安金融大厦消费欧陆风格不同，它消费的是民族传统风格，但殊途同归。更重要的是，操纵这种中国民族风格消费的却是来自美国的连锁集团。资本的力量早就跨越国界，也无视文化之间的差异——在消费社会里，资本只关心表面上迎合大众趣味，以及背地里的利润攫取。

This Tibet Mansion, which puts on a big hat and a set of armor, seems to condense the whole Potala Palace into one single building. In contrast to the pseudo-European style adopted by Ping'an Finance Mansion, what this building consumes is traditional Tibetan ethnic style. However, different ways lead to a same purpose — to consume. What's more, it is the hotel chain from the USA that initiated the consumption of such a traditional Chinese ethnic style. Capital has crossed national boundaries ignoring the variety in different cultures. In such a consumer society, capital is concerned with nothing but catering vulgar tastes apparently but seizing profits secretly.

场所：黄浦区人民路、河南南路交界处的西北角
功能：商铺＋旅店＋住宅
Site: the north-western corner of the intersection of Renmin Road and South Henan Road, Huangpu District
Function: shop + hostel + private dwelling

07 楔形屋
Wedge-shaped Building

从总平面上看这栋房子就如同一块楔子，这种形态由互成锐角的河南南路与金门路构成。在这栋楔形屋中，好几种功能集中在一起：底层一圈为商业店铺，招待所与小旅馆位于中部；顶层是一户户带有屋顶露天平台的私人住宅，这些住户基本都是租客，过着早出晚归的生活。

The plan of this building is similar to the shape of a wedge, due to the acute angle formed by South Henan Road and Jinmen Road. Various functions are merged together in this Wedge-shaped Building. Shops take the ground floor, and small hostels and inns are at the intermediate level. On the roof are all personal houses with open-air terraces. Local inhabitants are almost tenants, who lead a life of going out at dawn and coming back at dusk.

中国工商银行
人民路
W
E

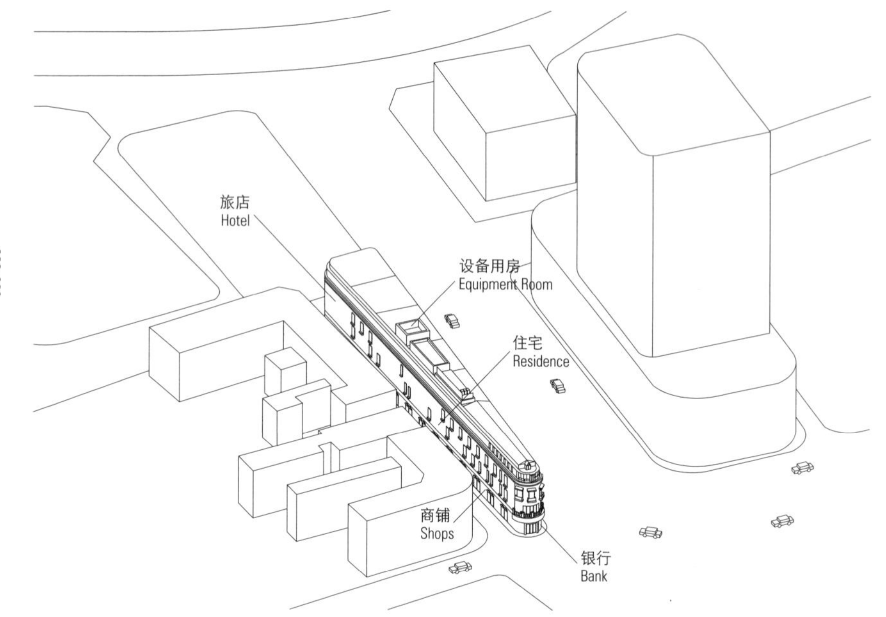

上海有一种有趣的现象，城市道路将一个地块在面积上进行细分的同时，地块内的建筑也通过细分及转让实现多种功能的共生。然而，与土地私有的东京在遗产税的影响下导致七零八落的剩余地块不同，在土地国有的情况下，上海的地块随时可能因为建筑过于破败而遭强拆。楔形屋是这股浪潮中的幸存儿。它的形态在城市空间中是孤立的，而它自下而上从商铺到旅店，再到住宅的功能递变，也反映出与城市关系由密到疏的变化。

An interesting phenomenon in Shanghai is that, when a block is subdivided by its nearby roads, the coexistence of several functions is realized through subdivision and transformation of the buildings within the block. However, compared with the fragmental in-between blocks influenced by the inheritance tax in Tokyo, where a private ownership of land exists, buildings of poor quality in Shanghai may be compelled into demolition at any time, since the ownership of land belongs to the state in China. The Wedge-shaped Building is a lucky survivor in such a trend. It has an isolated form in this urban space, and the transmutation of the functions from bottom up — from shops to hostels, and to private dwellings — reflects that its relationship with the city changes from open intimacy to privacy.

馆

第二皮鞋厂
中国工商银行

场所：闸北区莫干山路 50 号
功能：工地 + 码头 + 酒店 + 创意园区
Site: No.50 Moganshan Road, Zhabei District
Function: construction site + wharf + hotel + creative park

08 废墟游乐场
Playground in the Ruins

从黄浦江沿苏州河逆流而上，会遇到很多处急弯，在第一个超过 90° 的大转弯旁，有一片荒废已久的园区。但这里并不死寂，住在周边的孩童常常在废墟中玩耍，很多摄影发烧友慕名前来取景。在这片废墟游乐场里，遗存房子的外立面几乎都被绘上狂放的涂鸦。废墟旁边是作为苏州河游船线路上下客点之一的昌化路码头。

When going upstream from Huangpu River along Suzhou Creek, one will encounter many sharp turns. Beside the first sharp, there is an area that has been abandoned for a long time. However it isn't deadly silent here. The children living nearby often play in the ruins, and a great many photography hobbyists come to take pictures in pursuit for these desolate scenarios. Within the site, all the external walls of the remaining buildings were painted with graffiti. Changhualu wharf, which is one stop of the cruise line of Suzhou Creek, is just beside the ruins.

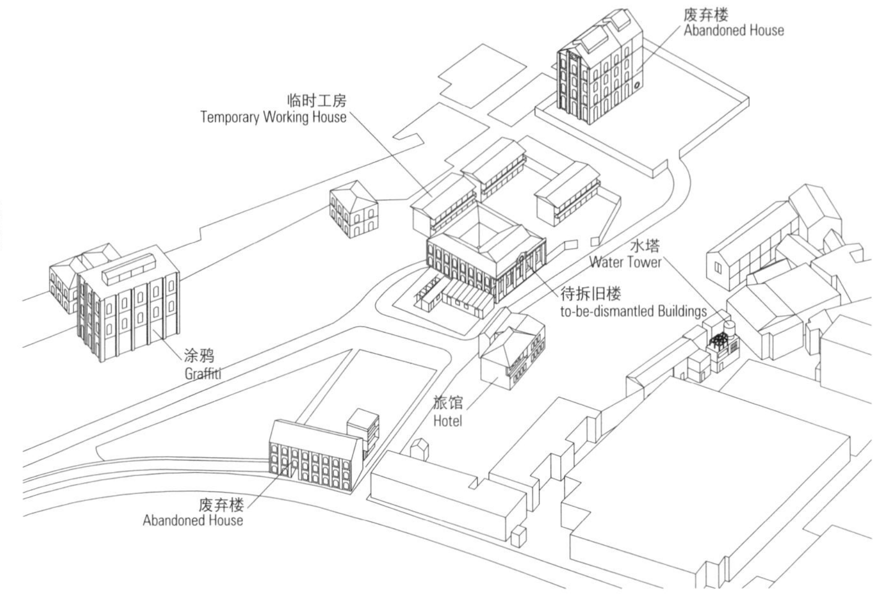

在上海，民族工业时期的很多老厂房现在都已被推倒，它们的历史残片却成为现代人娱乐的场所。废墟游乐场作为这种荒废场地的典型，它的未来很不确定。在城市急剧乃至畸形发展的当下，也许某天我们回访废墟游乐场时，意外地发现这批现在遗存着的房子都被一拆而光。上海的这类废墟应当得到更多世人的关注与记录，即便无力挽回它们的形体，也应存下一份记忆。

In Shanghai, a lot of former factories of the national industry period are being torn down, while their historical fragments become an entertainment place for modern man. Being a typical case of such disserted fields, this playground in the ruins has an uncertain future. In the rapid and even abnormal urban development today, when we pay a return visit to this case one day, we may be astonished to find that all these remaining buildings have totally been dismantled. We really hope that such ruins in Shanghai can concern the public and be recorded by more and more people. The memory must be kept, even though the ruins may not survive in the coming future.

场所：黄浦区虎丘路 20 号
功能：展览
Site: No.20 Huqiu Road, Huangpu District
Function: exhibition

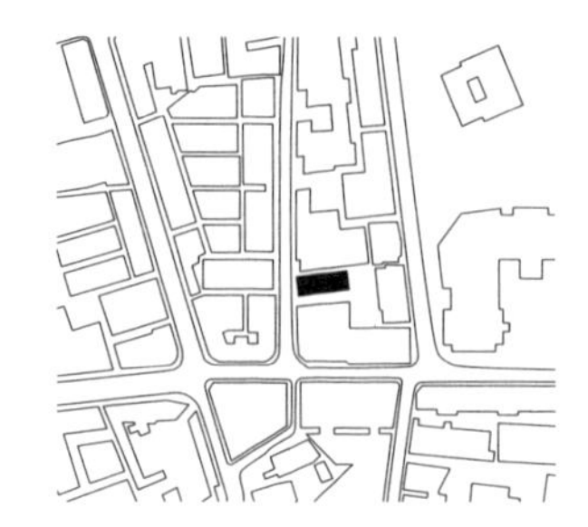

09 外滩美术馆 Rockbund Art Museum

外滩美术馆位于上海外滩源，其前身是亚洲文会大楼，由英国皇家亚洲文会北中国支会出资，建成于 1932 年，内设图书馆、博物院和演讲厅。建筑采用装饰派艺术风格，外立面上使用精致的水刷石做法。2007 年，美术馆邀请英国建筑师大卫·奇普菲尔德主持室内改造。他用一个三层通高的新中庭将四至六层连接起来，顶棚上开天窗，可以更有效地利用自然光。

Rockbund Art Museum is located at the center of Bund Origin of Shanghai. Formerly it was the building of Royal Asiatic Society, sponsored by North China Branch of the Royal Asiatic Society and completed in 1932. At that time there was a library, a museum and a lyceum inside the building. It adopted the Art Deco style, and utilized Shanghai plaster, a kind of exquisite material, on the façade. The British architect David Chipperfield was invited to direct the interior renovation project in 2007. He integrated the 4th, 5th and 6th floors by adding a new central hall spanning the full 3-story height. Skylights are brought in through the roof windows.

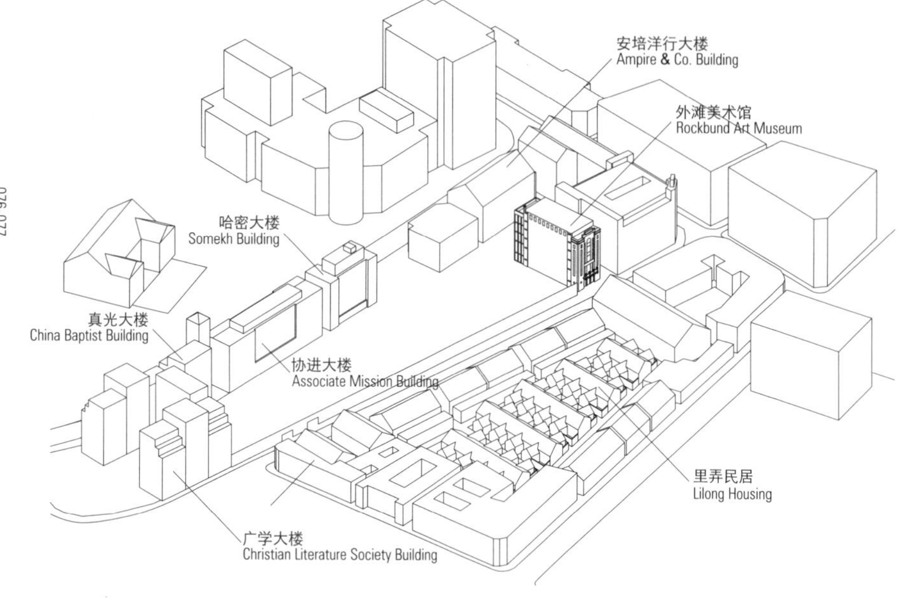

外滩美术馆是上海 20 世纪 30 年代那批采用装饰派艺术风格的建筑中得以延续的优秀例子。现在，它的功能由原先的图书馆与博物馆被置换为类似的美术馆，实现了一次非常成功的转型。如今，外滩美术馆会定期邀请各类艺术家和建筑师做展览或讲座，将把艺术、媒体、城市等领域的探讨呈现给公众。

Rockbund Art Museum, which has achieved an afterlife, is an excellent example of all those Art-Deco architectures during the 1930s in Shanghai. Nowadays its function has switched from the original library and museum institute to an art museum, realizing a very successful transformation. Nowadays, Rockbund Art Museum regularly invites all kinds of artists and architects to hold exhibitions or give speeches, through which discussions in various fields such as art, media and city are presented to public.

场所：虹口区梧州路 384 号
功能：餐馆 + 住宅 + 鸽舍
Site: No.384 Wuzhou Road, Hongkou District
Function: restaurant + residence + pigeon house

10 都市巢居 Urban Nest

梧州路两侧大多是细碎、拼贴化的城市肌理，“都市巢居”是其中功能与空间混搭的典型。它总共五层高，与两侧两层高的房子紧贴在一起。三种迥然不同的功能堆积木般排列组合：底下两层是服务周边人群的餐厅；第三、第四层是用于私密居家生活的住宅；顶部后加出的阁楼空间用作鸽舍。

The city patterns on both sides of Wuzhou Road are mostly fragmental and collaged. Among the collage is which the building nicknamed "Urban Nest" is the typical one of mixed functions and spaces. It totally has five floors and is closely connected to the bilateral houses. Three different kinds of functions are stacked together like toy blocks. The lowest two floors are used as a restaurant for the people living nearby, while the second and third floors are for private domestic living. An attic space used for housing pigeons was added later on the top.

北国村
东北炒菜
手工水饺
小四川
炒
菜

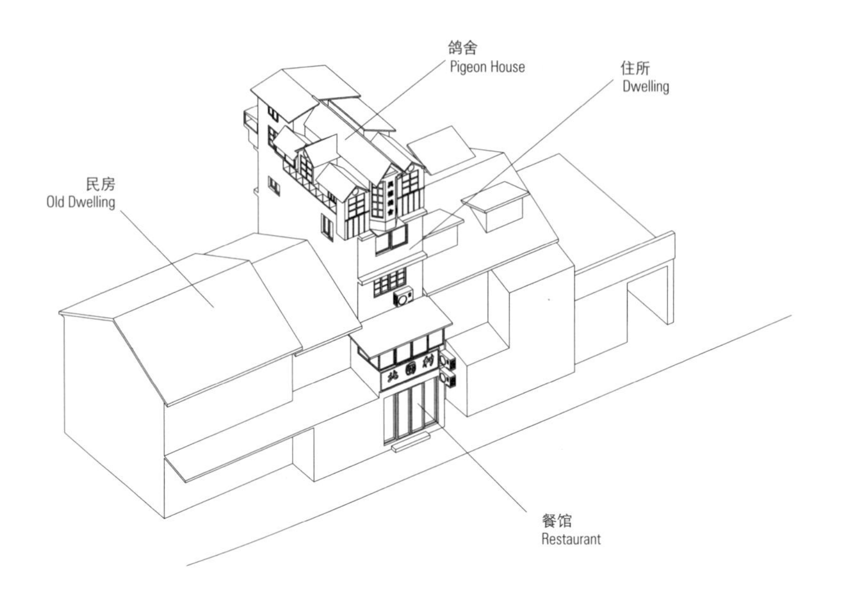

“都市巢居”是日常建筑将功能杂交的典型。户主居住于中间层，底层经营餐厅以养家，顶层饲养比赛用鸽，并作为生活中的乐趣与寄托。底层民众往往具有天才的空间改造力，孕育出建筑师在一栋正统的建筑上考虑不到的各种可能性。或许这就是所谓的“滥建筑”带给我们这个都市最大的启示。

Urban Nest is a typical case of everyday architecture with hybridized functions. The owner lives in the middle, earns his living by operating a restaurant beneath, and rears his pigeons for racing games on the top. Common people have the most ingenious talent to create any possible spaces, compared with the formal architecture by an architect. Perhaps it is the most valuable inspiration which the so-called "da-me" architecture (or "not-good" architecture) brings to our city.

IVE
上海市汽车修理有限
柴油汽车修

场所：杨浦区隆昌路 378 号
功能：居住
Site: No.378 Longchang Road, Yangpu District
Function: residence

11 猪笼寨 Walled City

1932 年（民国二十一年），公共租界工部局在隆昌路 362 号格兰路巡捕房北侧建钢筋混凝土 5 层楼公寓 1 座。做巡捕房用，新中国成立后改为公安局职工宿舍，现为普通居住用房。建筑面积 10 200 平方米。中有天井，四周密布住所，每间 19 平方米。二层起走廊环通，并配有电梯 1 部。原本宽敞的走道上布满了厕所、储物间、洗手台等加建体，几户人家往往共用这样的空间。围合起来的天井常被用作晾晒空间，还设有专门的垃圾回收点。

In 1932 the Municipal Committee of public concession built a 5-story concrete apartment at No.362 Longchang Road. It was used for the police before the founding of PRC, and then changed into a dormitory for the staffs in Public Security Bureau. Now used as a normal residence, this building covers an area of 10,200 square meters and has a middle courtyard, surrounded by densely arrayed rooms, each of which occupies 19 square meters. There are circulated corridors above the ground floor, and one elevator. Newly-added toilets, closets and washbasins almost block the wide passage, and they are usually shared by several families.

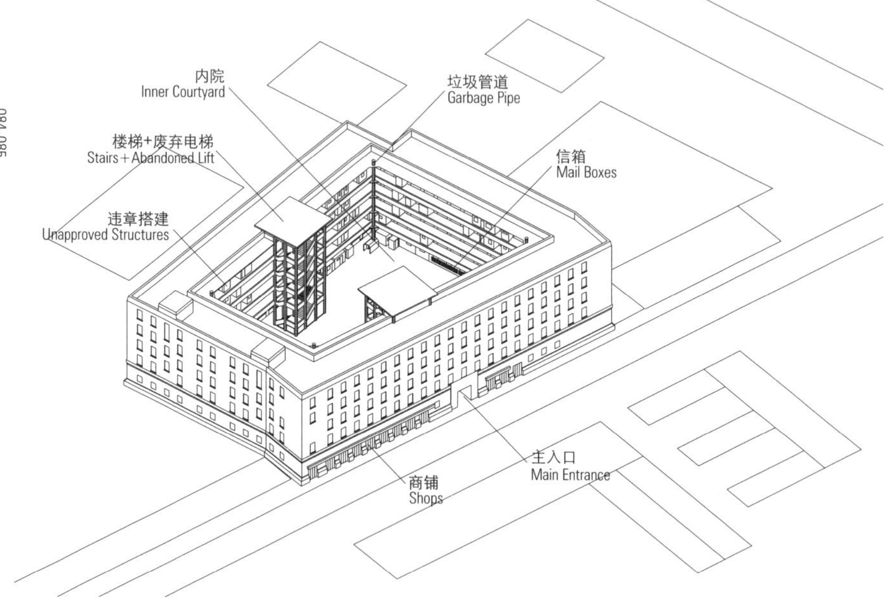

“猪笼寨”这个名字源自这栋公寓四面围合的“口字形”布局，以及内部高度密集的生活社区。通过一个不起眼的大门拐进这个别有洞天的世界，会发现在小小的院落里，人们自由自在地生活，形成一个与外界隔绝的社区。

The name "Walled City" comes from the enclosed layout of this apartment building and the community of high density inside. Passing through an unimpressive gateway, one can discover an isolated world, in which local inhabitants are living together freely and peacefully.

场所：天目西路南北高架底下
功能：菜市场 + 运动设施 + 商店 + 停车 + 仓储
Site: beneath the South-North Viaduct on Tianmuxi Road
Function: market + sports facilities + shop + parking lot + warehouse

12 高架综合体 I
The Complex under an Elevated Road I

天目西路南北高架上是川流不息的车流，底下的剩余空间则被利用起来，整合为具备停车、菜市场、仓储、运动等多种功能的综合空间。这种综合空间并非只是底层市民的自发行为，还包括当地街道办的大力组织。当年建造菜市场时街道办甚至还请过设计单位绘制图纸。该菜市场具有双层屋顶，外层为高架桥的桥底，内层由钢板与塑料组成。这里的篮球场、乒乓球场等运动设施的使用效率都比较高。

There are streams of cars on the South-North Elevated Road above Tianmuxi Road, while the space under it was used for a multi-functional complex including parking lots, a market, warehouses and sports facilities. This complex wasn't merely the result of the spontaneous behavior of local citizens, but due to the effort by the local street office. They even asked a design studio to make the drawings of the market, which has a double roof, with one layer being the bottom of the elevated road and the other being constituted of steel plates and plastics. The facilities such as the basketball and table tennis fields all have high service efficiency.

天目西菜市场
Tian Mu Xi Cai Shi Chang

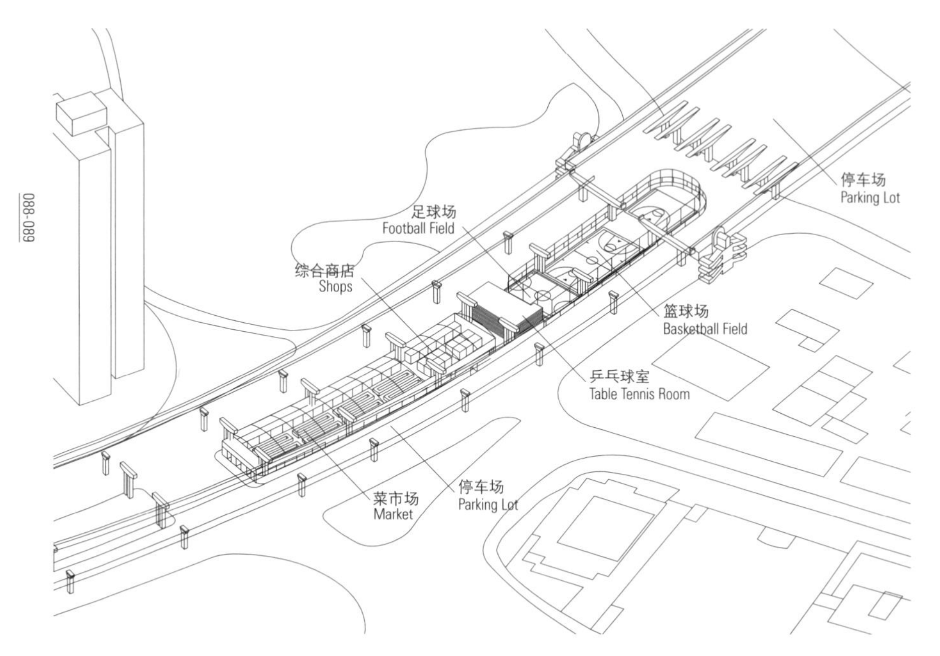

本书一共有两个位于高架路底下的综合体案例，这是第一个。这些功能共生于同一个遮蔽空间下，相互之间关联不大，却都有同一目的——最大化地利用剩余空间。另外，街道办作为政府基层部门介入到这类空间的改造与利用中，说明自上而下的“设计”与自下而上的“非设计”可以找到某一个契合点。

In this guidebook there are two cases of the complexes under an elevated road, this being the first one. These functions coexist under the same shelter, and even though they barely have any relations with each other, they all share the same purpose, which is to make the most of in-between spaces. In addition, the role played by local street offices as a grass-root government department into the renovation and utilization of such spaces indicates that the top-down "design" and the bottom-up "non-design" can reach an integration point.

火灾隐患,请拨打"96119"

场所：长宁区近虹桥路
功能：机场 + 火车站 + 地铁 + 公交 + 的士 + 酒店 + 商业
Site: near Hongqiao Road in Changning District
Function: airport + railway + metro + bus + taxi + hotel + commerce

13 虹桥枢纽 Hongqiao Complex

为迎接世博会，虹桥枢纽全面建成于 2010 年，包括高铁站房、磁浮车站、地铁车站、航空机场和长途车站等设施，所形成的高度集成的超大型城市轨道综合体，是上海交通最重要的节点之一。它通过整合公路、铁路和航空三大交通方式，每天日夜不断地输送各类客流与货流。

In order to serve the Expo, Hongqiao Complex was completed in 2010. It includes facilities such as the high speed train station, the maglev station, the metro station, the airport and long-distance bus line terminal. This highly integrated supercomplex of urban rails is one of the most important transportation nodes in Shanghai. It merges the three main kinds of transportation — highroads, railways and aviation. All kinds of flows of passengers and goods are moving through the complex day and night.

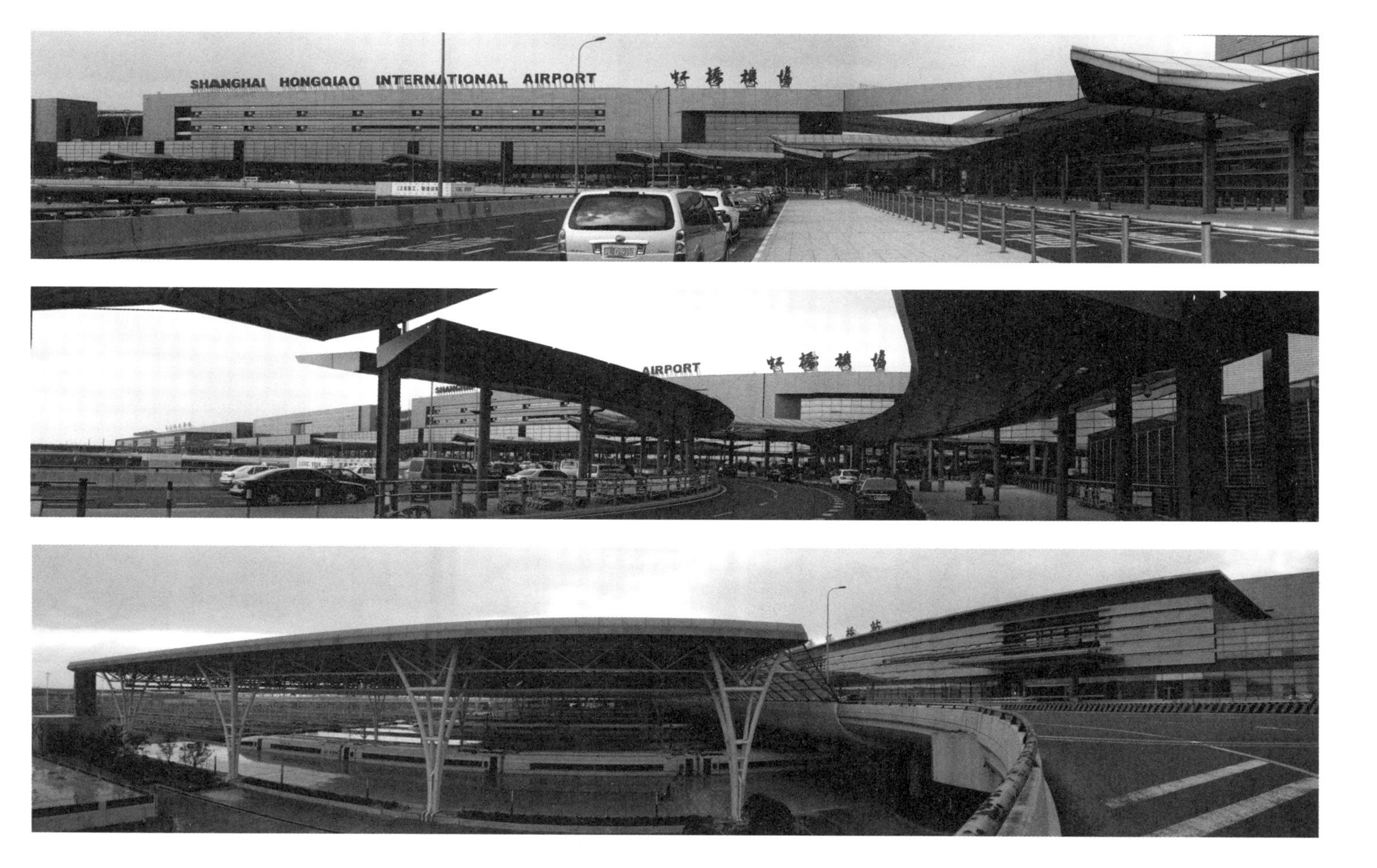
SHANGHAI HONGQIAO INTERNATIONAL AIRPORT
AIRPORT

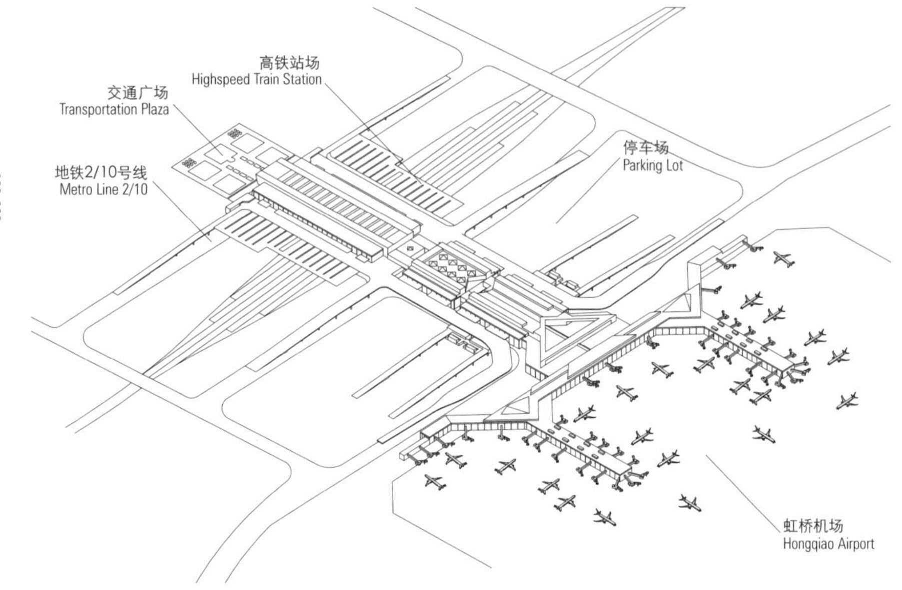

作为超大型交通网汇集点的虹桥枢纽已成为上海的门户，因为很多旅游者必然通过这里进入城区。高度发达的技术文明组织多层次的交通类型于一体，这类项目反过来也最为直接地推动工业和信息产业的蓬勃。从这一点来看，上海已与东京、巴黎、纽约等大都会，并列为交通技术全球化的重要基地。

As a super transportation node, Hongqiao Complex becomes the portal of Shanghai, for many tourists enter the downtown by going through it. Highly developed technological culture organizes different layers of transportation together, creating a super infrastructure that directly promotes the prosperity of industry and information industry. From this viewpoint, Shanghai becomes one of the leading sites of transportation technology on the globe similar to the other metropolises such as Tokyo, Paris and New York.

AIRPORT

2F 出发层
Departures Floor
B1铁路售票处
Railway Ticket Office
2号航站楼
Terminal 2
1-15到达(北)
餐饮
1-15到达(南)

B1
P
VIP

8
10

场所：延安路高架与南北高架交接处
功能：交通
Site: intersection of Yan'wwan Elevated Road and South-north Elevated Road
Function: transportation

14 龙柱 Dragon Pillar

龙柱位于延安路高架与成都路高架的交汇处，形成复杂的六层高架系统。下有地铁通过，汇集了大量的人流与车流。之所以称为“龙柱”，是因为在原本只承担结构作用的钢筋混凝土柱子外雕刻了龙的纹样。层层叠叠的高架路面也如同巨龙，与龙柱上的腾龙纹样交相辉映。

Dragon Pillar is at the center of the intersection of Yan'an Elevated Road and Chengdu Elevated Road, making it a complicated 6-layer system. There is metro system beneath, adding large flows of people and cars converged at the location. It's called "Dragon Pillar" because dragon patterns are engraved onto the concrete pillar. Layers of the above elevated roads also appear like dragons. Their literal and figurative representation enhance each other's beauty.

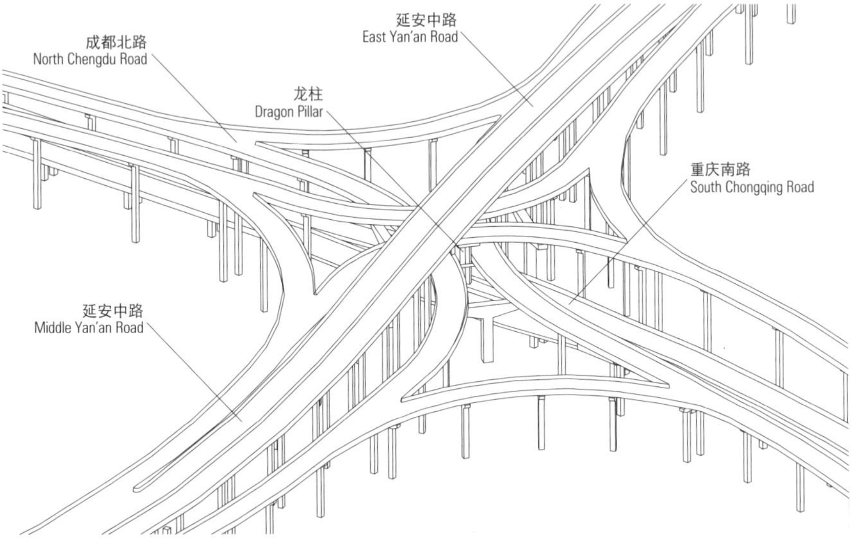

如何实现龙柱上方六层高架的依次叠合是一个技术难题。然而，龙柱本身还有一段离奇而玄秘的故事，这使得它超越一般意义上只具有纯粹功能的交通设施。柱子上环刻的龙浮雕所具有的象征与隐喻含义，显示出都市人对都市传奇的热衷与想象。

How to superimpose those 6 layers of elevated roads above the pillar is a logistical problem. However, the pillar itself has an odd story, making it transcend the ordinary mere functional transportation facilities. The symbol and metaphor in the dragon relief engraved around the pillar indicate urbanites' zeal and imagination of urban legends.

场所：黄浦区江西南路 20 号
功能：隧道通风 + 办公楼 + 商店 + 售票 + 卫生间 + 广告
Site: No.20 South Jiangxi Road, Huangpu District
Function: tunnel ventilation + office building + shop + tickets office + restroom + advertisement

15 通风塔 Ventilation Tower

隧道的通风塔并非只是具有单一功能的附属设施。在这栋独立建筑中，汇合了办公楼、商店、卫生间等另外几类功能，分布在附属空间中。这栋房屋顶部高挂着写有“上海欢迎您”的标语牌以及其他广告牌。

This tunnel ventilation tower doesn' t merely have one specific function. In this isolated building, several other functions are converged in the annexed spaces, such as offices, shops and WCs. The billboard saying "Welcome to Shanghai" and other advertisement boards are placed on the top of this building.

上海欢迎您

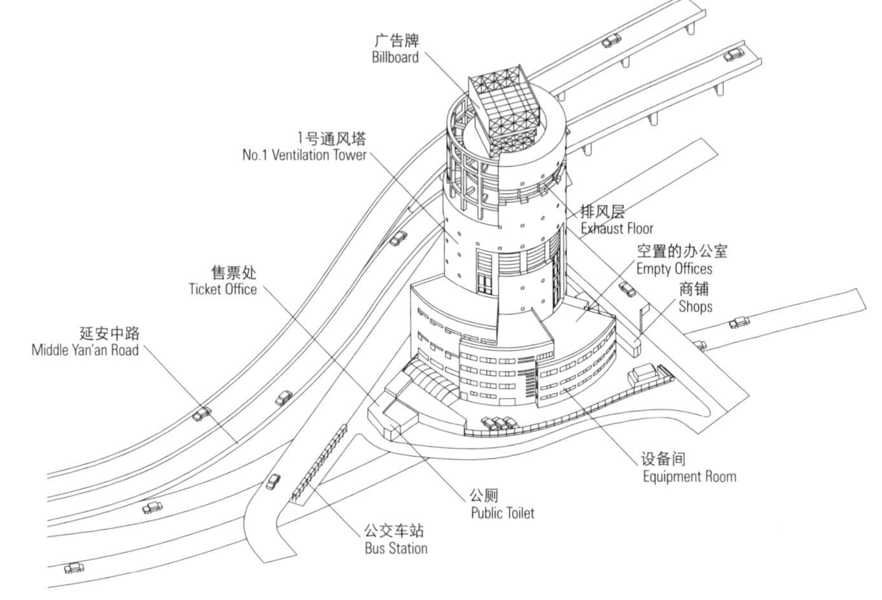

在本案例中，作为城市基础设施的通风塔衍生出一系列日常功能空间。通过与这些附属功能的组合，通风塔将它与周边环境的关系由原本消极的寄生形式转变为相对积极的共生形式。每项功能之间互不干扰，不过这种共生关系也意味着容易被另外的功能取代。

Ventilation Tower is a kind of urban infrastructure, and a series of spaces of everyday functions derive from it. By combining these annexed functions, Ventilation Tower reverses its relationship with the surrounding environment from parasitizing to coexisting, namely from a negative way to a positive one. No interference is among these functions, however, such a relationship also indicates that it is easy to replace any one of them.

场所：南浦大桥西侧
功能：交通
Site: north of Nanpu Major Bridge
Function: transportation

16 南浦大桥盘旋高架
Spiral Viaduct of Nanpu Major Bridge

南浦大桥盘旋高架是内环线上的一个重要节点和大型交通枢纽。巨大的盘旋坡道与错综复杂的连接使其成为该地区最主要的地标。盘旋高架下的空间容纳了一个公交车站与一些公共绿地空间。

Acting as a traffic hinge, the Spiral Viaduct of Nanpu Major Bridge is an important node of the Inner Ring Road. Giant spiral slopes and complicated connections make it the landmark in this district. The space beneath it contains a bus terminal and some public green spaces.

停车场
南京银行

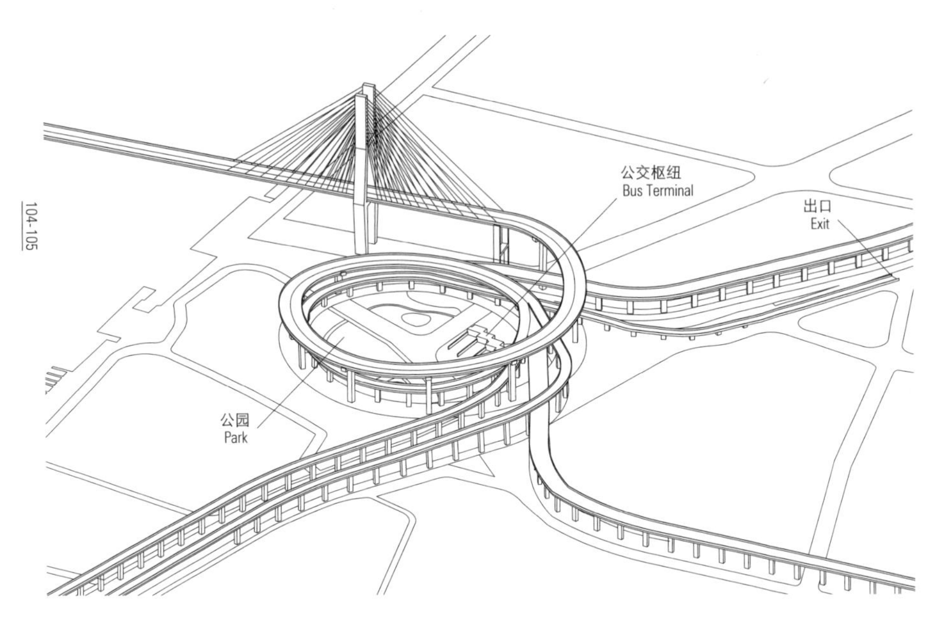

南浦大桥巨大的尺度来自两方面，一是为黄浦江江面提供足够的航道净空，二是保证黄浦江两岸的交通联系。这类大都市基础设施的巨构尺度在推动路桥结构技术发展之余，也足以为都市空间的利用带来更多的可能。假如桥底下的空间能获得更充分的利用，那么也许一个将剩余空间发挥到极致的“超级南浦大桥”会出现。

There are two factors that lead to the giant scale of Nanpu Major Viaduct, one to provide enough net height for the channel on Huangpu River, and the other to ensure the relations of traffic on two sides. Such kind of metropolitan infrastructure not only pushes the development of the structural technologies in roads and bridges, but also creates greater possibilities in the utilization of urban spaces. If the space beneath the viaduct could be made more advantage of, a "Nanpu Super Bridge" making the most of in-between spaces may come into being.

场所：卢湾区建国中路 8-10 号
功能：创意园区 + 商业 + 休闲娱乐
Site: No.8-10 Middle Jiangguo Road, Luwan District
Function: creative park + commerce + leisure and entertainment

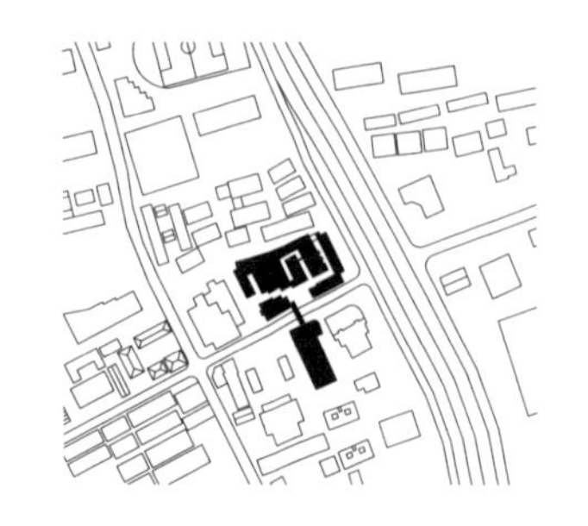

17 八号桥 Bridge Eight

八号桥占地面积 7 000 多平方米，总建筑面积约 1.2 万平方米。这里曾是旧属法租界的一片旧厂房，新中国成立后成为上海汽车制动器公司所在地。进入 21 世纪后，由于原企业重组后搬迁，留下了这七栋旧厂房，现在被改造成为创意园区。天桥连接起散落的体量，形成独具标示性的入口。同时，天井、庭院、平台等公共空间被精心地组织起来，形成一个办公、商业、旅游等功能并置的场地。

Bridge Eight takes an area of about 7,000 square meters, and has a floor area of 12,000 square meters. Originally it was a group of old factories in French Concession, and was turned into Shanghai Brake System Company after the founding of PRC. Upon the arrival of the new millennium, the company moved out because of rearrangement, leaving these old factories vacant. Now they have been renovated into a creative park. A bridge links the scattered volumes together, forming a conspicuous entrance. Public spaces such as courtyards and terraces are organized to make a site with various functions including offices and commerce.

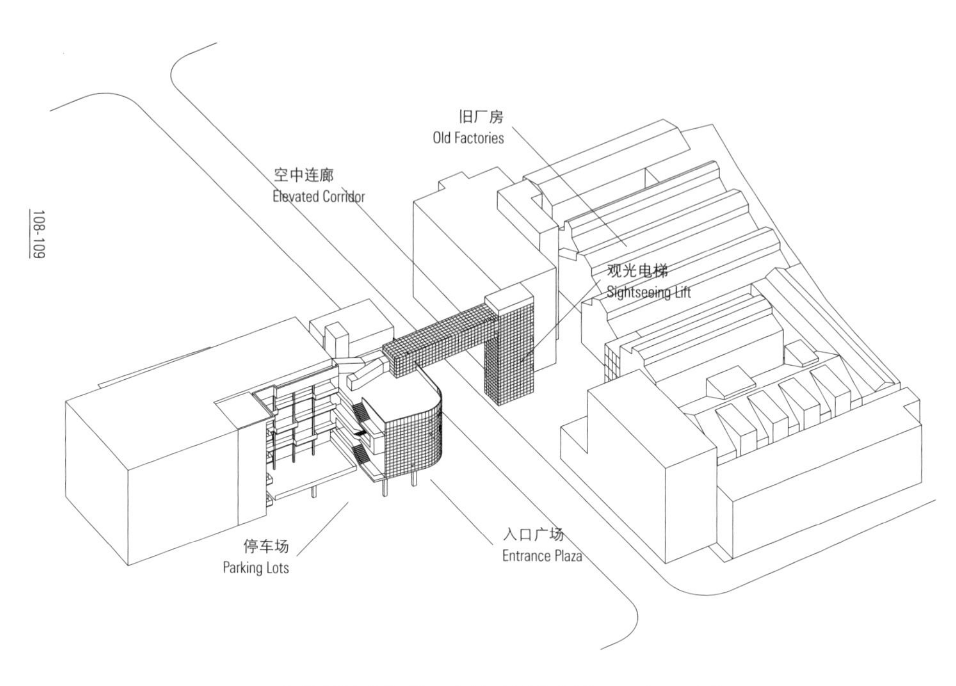

八号桥是上海老厂房被改造为创意园区的一个典型且成功的案例。它清晰地展现出如何在原有的空间躯体中实现功能置换，如何通过空间改造的方式刺激出一个场所的能量。

Bridge Eight is a typical and successful case of turning an old factory into a creative park in Shanghai. It clearly exhibits how the function is replaced within the original space, and how the energy of a place can be stimulated through the method of space renovation.

喧哗 自有声
新君越 全系上市
北京现代

场所：黄浦区宁波路 400 号
功能：商业 + 住宅
Site: No.400 Ningbo Road, Huangpu District
Function: commerce + residence

18 兵舰头
Warship-head-shaped House

这栋房子所处的区域保留大量四方形的老旧民居，而它与众不同的是，其平面并非规整的方形，而是东面斜着朝内转折，因此它与东侧的建筑形成一条弯折而狭窄的巷道。在宁波路由东往西看，它的外形酷似舰船的头部，于是被昵称为“兵舰头”。

The region in which this house is located retains a large number of old rectangular residences. However, this house differentiates from all its neighbors. It curves inward alongside the east interface, resulting in a meandering and narrow alley on the east. Its external shape seems very much like the head of a warship when viewing westwards from Ningbo Road, and this explains why it is nicknamed a "Warship-head-shaped House" .

辰龙商务酒店
TCL
legrand
罗格朗

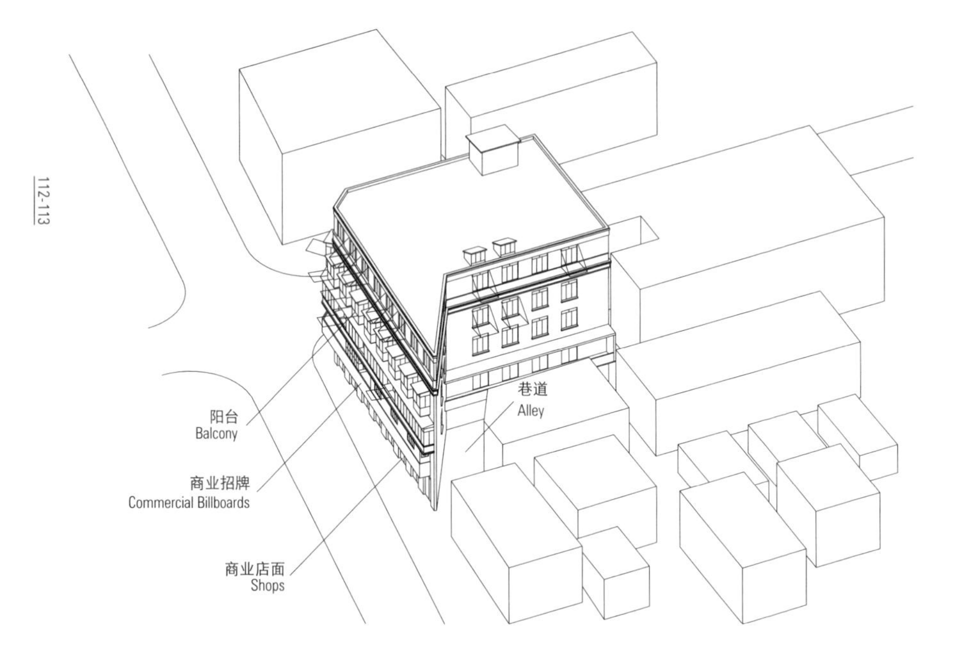

两条道路相交时形成的锋利锐角造就了这栋房子尖锐的外观，也就是说，宏观的城市格局决定了微观的建筑形态。这种道路的出现可能是由于地块的细分，或者是原有地形地貌中有某样因素的影响。如此尖锐的建筑轮廓必定影响室内布局，甚至影响到住户的日常行为——城市街块的变更会投射到人的一举一动上。

It is the sharp acute angle created by the intersecting of two roads that makes the sharp external shape of this house. In other words, the layout of the macro urban environment determines the micro shape of a building. Such roads are perhaps due to the subdivision of blocks, or a certain factor in the original topography. Now such acute external silhouette influences the interior, and even influences the behaviors of the inhabitants — the alterations of the urban blocks may project into the movement of human beings.

有限公司
销商 电记

场所：黄浦区新闸路 258 号
功能：停车
Site: No.258 Xinzha Road, Huangpu District
Function: car parking

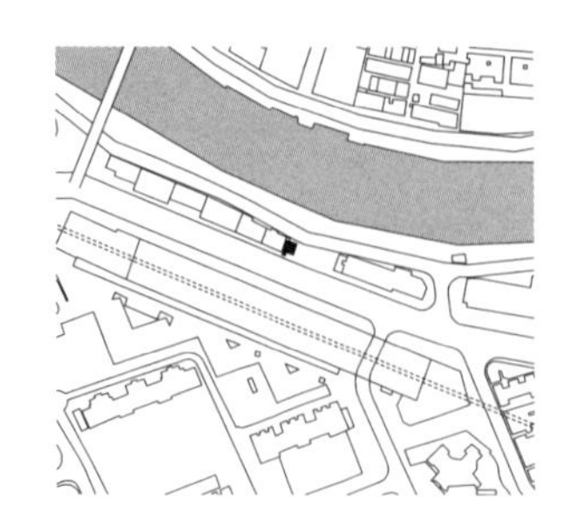

19 停车塔 Parking Tower

停车塔占地 81 平方米，能够容纳 38 辆车，是立体式停车空间的典型。停车塔内利用液压起重机上下运输车辆，外观上用粉红色与白色交错的喷漆铝板塑造了一个纯净的长方体。在苏州河南岸的这片高楼建筑群中，它虽然不高，但很醒目。

Covering only 81 square meters and accommodating 38 cars, Parking Tower is a typical case of vertical parking. A hydraulic crane is used to raise the cars within the building, while painted pink and white aluminum panels on the outside give it a pure cuboid form. It is not very tall among all the other high-rises that are alongside the south shore of Suzhou Creek, however, its unique stature makes it quite conspicuous.

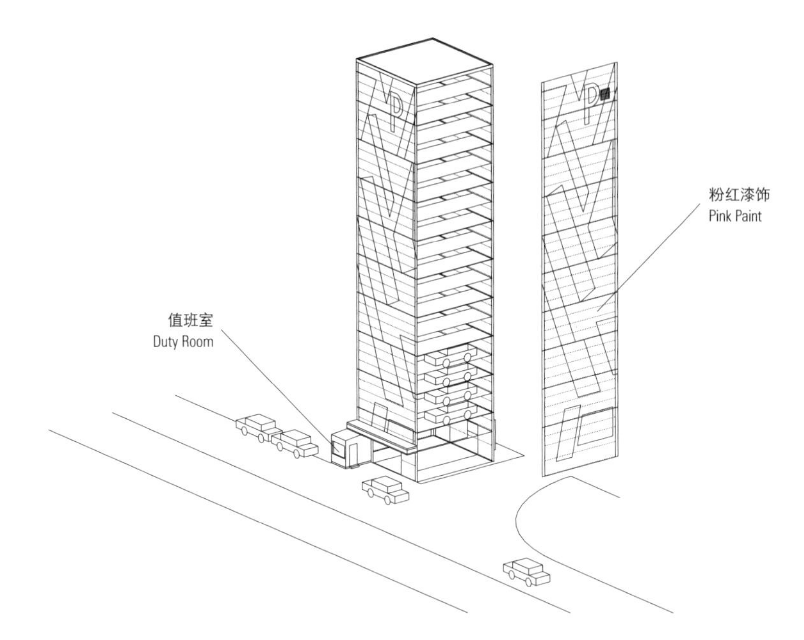

大型地下停车库原本是提供城市停车位的首选，然而停车塔却因为地权的细小划分以及高密度的需求而出现了。人口激增导致城市高容积率区块的产生，同样，停车塔是车的世界里高容积率的典型代表。车辆的垂直运输需要大型机械装置来辅助完成。我们会发现，机械化在城市中牢牢地掌控一切，构建起我们的高密度生活。

Originally, an underground car parking lot is considered as the first option for parking, but but as density increases and land properties are subdivided a new method of parking was developed. The fast development of population causes high-FAR (Floor Area Ratio) urban blocks. This is similar in the world of cars, in which, Parking Tower is a great example of high-FAR parking structures. The vertical movement of cars requires large mechanical installations in it. As seen in Parking Tower, it can also be noticed that mechanization firmly takes command in the city and fabricates the high-density life.

黄河路立体停车库
24
小时 停车

场所：黄浦区西藏南路 1 号
功能：娱乐 + 商业
Site: No.1 South Xizang Road, Huangpu District
Function: entertainment + commerce

20 大世界 The Great World

大世界始建于 1917 年，曾是上海最大的室内游乐场，素以游艺、杂耍、南北戏剧和曲艺为特色，是上海最具代表性的娱乐建筑。1924 年重建为钢筋混凝土建筑，现占地面积 6 500 平方米，建筑面积 13 580 平方米。内部有游艺厅、歌舞厅、魔奇世界、桌球房、乒乓房、滚轴溜冰场等数十项娱乐活动项目的场所，并有中式餐厅和上海特色小吃廊。

The Great World was founded in 1917, where it was once both the largest indoor amusement park and the most influential entertainment architecture in Shanghai. It is famous for vaudevilles and Chinese traditional folk operas. In 1924, it was rebuilt to occupy an area of 6,500 square meters and have a floor area of 13,580 square meters. Within the new design there were various recreational spaces, such as, dancing halls, magic-world halls, billiard halls, ping-pong halls, and roller-skating rinks. Furthermore, Chinese restaurants and snack bars for Shanghai's delicacies could also be found in the 1924 Great World.

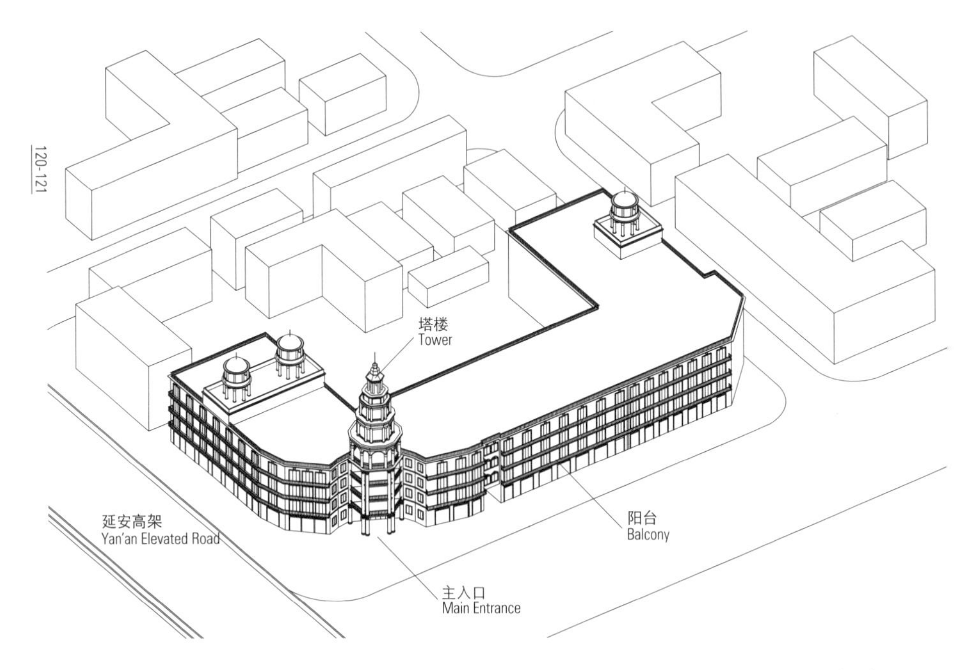

大世界是中国早期具有主题公园性质的场所，将多种娱乐设施连同配套的休闲设施整合到一个大空间里。作为 20 世纪 20 年代摩登上海充满戏剧化的一面，大世界编织、疏离并且重构现实，在有限的空间中为观众呈现出一个无边境的“大”世界。

The Great World, which had integrated various entertaining and supporting facilities into a single large space, was one of the earliest amusement parks in China. As a dramatic face of Modern Shanghai in the 1920s, the Great World had woven, alienated and reconstructed reality; thus, creating an infinite "great" world to its audience within a limited space.

场所：徐汇区新乐路 167 号（东湖路交接处）
功能：餐饮
Site: No.167 Xinle Road, Xuhui District
(at the intersection with Donghu Road)
Function: restaurant

21 假山餐厅 Rockery Restaurant

假山餐厅位于街道的转角，钢框架玻璃幕墙与片状金属屋顶包裹着一个矗立于假山之上的中式亭子。室内的木地板高出室外地坪约 80 厘米，围绕着假山布置有一圈餐桌。该餐厅主打日式餐饮，比如日式拉面、定食、咖喱等。

Rockery restaurant is located at the intersection of two streets. Centered on the roof top made of metal panels, stands a Chinese style pavilion; while surrounding the restaurant is a modern glass curtain wall in steel frame. Raised approximately 80 cm higher than the outdoor ground, the interior wooden floor is lined with tables circled around a traditional rockery garden. The architectural setting complements with the Japanese food (stretched noodles, set meals and curry) served at the Rockery Restaurant in a unique way.

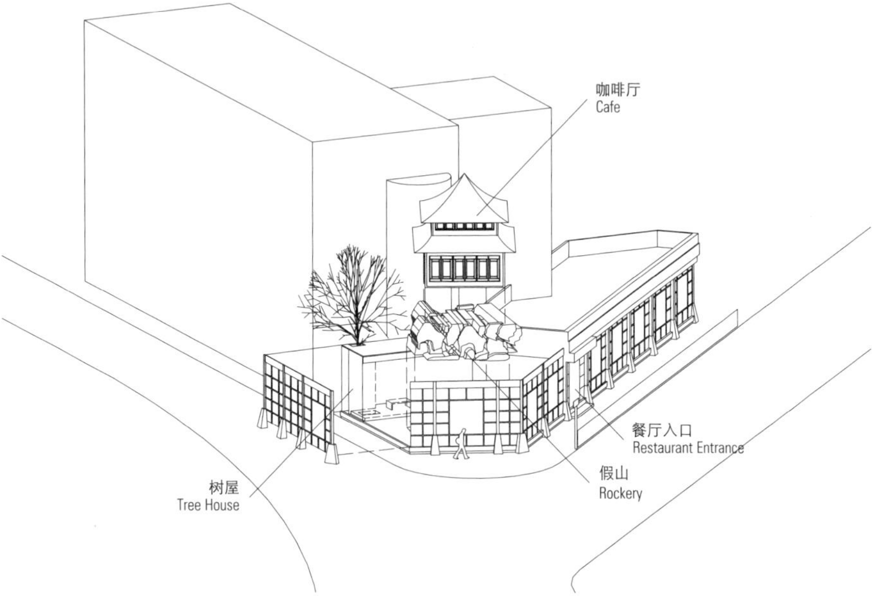

咖啡厅
Cafe

餐厅入口
Restaurant Entrance

假山
Rockery

树屋
Tree House

在一栋现代的钢与玻璃材料构成的房子里，置入传统中国园林中的假山与凉亭，两者形成的空间张力是本案例的最大特色。而内部经营的却是日式餐饮，可见形式与功能、传统与现代的界限可以随时被打破。

The spatial tension, which comes from the juxtaposition of a modern steel-and-glass house and the elements inside a traditional Chinese garden, such as a rockery and a pavilion, is the most striking character of this case. The tension continues, as the restaurant sells Japanese food in a building with traditional Chinese elements located in China itself. It seems that the boundaries between forms and function, or between traditions and moderns, can be broken at any time.

场所：南京西路 1686 号
功能：寺庙 + 商业 + 地铁 + 停车
Site: No.1686 West Nanjing Road
Function: temple + commerce + metro + parking

22 静安寺 Jing'an Temple

著名的江南古刹静安寺已有近 1 800 年的历史，是中国内地最重要的密宗道场。它现在占地约 2 万平方米，内有天王殿、大雄宝殿等佛寺标准配置。寺庙底下有停车场以及途经的地铁 2 号线，是一个大型的交通枢纽。

As a well-known ancient temple in Jiangnan Area, Jing'an Temple has a history of over 1,800 years, making itself the most important ashram of Esoteric Buddhism in China mainland. The temple has an area of about twenty thousand square meters, resembling a standard configuration of a Buddhist temple, including the Hall of Heavenly Kings and the Great Buddha's Hall. There are also parking lots and Metro Line 2 located beneath the temple, making it a large transportation junction.

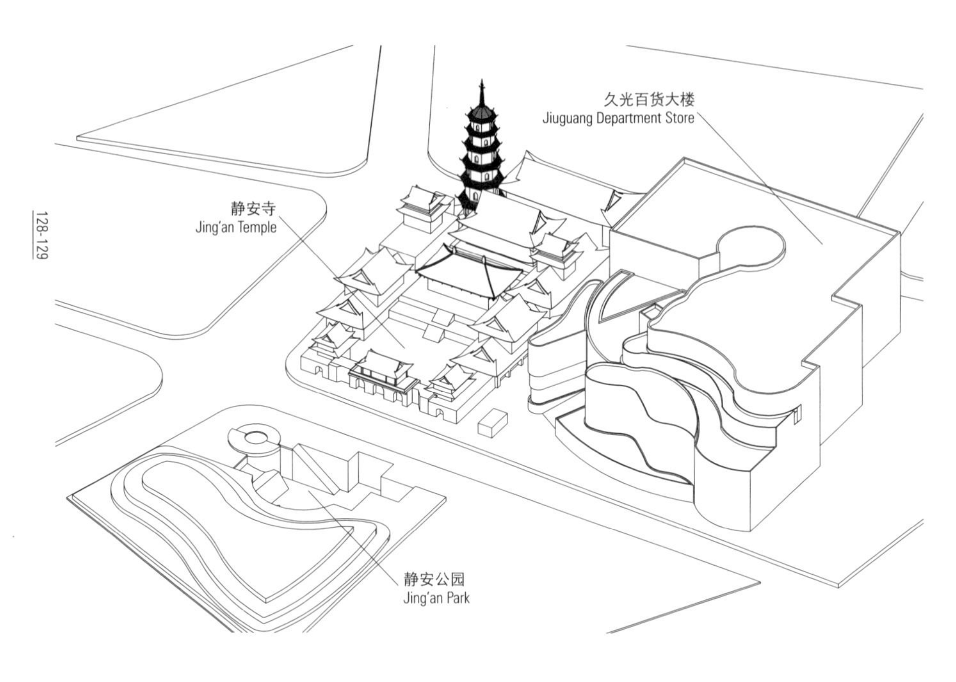

在文化与宗教占主导的传统地带，人流的聚集带来商业的蓬勃发展。然而，神圣与世俗间发生的纠纷也成为上海当下快速城市化进程中的怪现象。静安寺旁百货大楼立面上的巨幅广告牌与寺庙共生于一个时空里。消费社会的泛滥图像包围着曾经的禅修胜地，一切出世、永恒似乎都不可避免地卷入世俗的物欲洪流中，成为被后现代性肢解的对象。

In the traditional area where culture and religion take command, the flow of humans gathering brings the prosperity of commerce. However, the dispute of the sacred and the profane unexpectedly becomes a strange phenomenon in the rapid urban development of Shanghai. The gigantic billboards on the external surface of the nearby department store coexist with the temple. This past famous scenic spot of meditation is now being surrounded by all sorts of images of the consumer society. It seems inevitable that all the sacred and religious have been swept into the floods of fetishism, becoming the objects dismembered by post-modernity.

正法久住

般若

场所：黄浦江西岸
功能：亲水平台＋公园＋商铺
Site: west Huangpu Riverside
Function: waterfront platform + park + shop

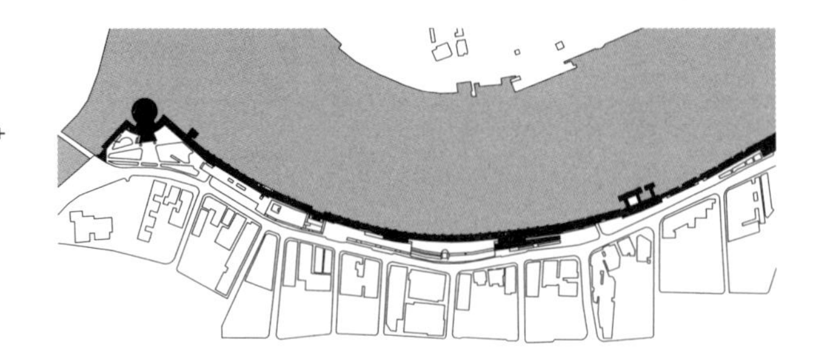

23 外滩景观大道
Sightseeing Avenue of the Bund

外滩是上海租界区也是整个上海近代城市开发的起点。景观大道长近 2 公里，它位于高出城市地坪约 8 米的平台上，首层有餐饮设施及外滩轮渡，底下是外滩隧道。外滩景观大道上最有名的景点是北端的人民英雄纪念碑、中部的外滩气象台以及南端的老码头。

The Bund is the origin of the urban development of the concession and modern city in Shanghai. It has a total length of about 2 kilometers, and sits on a long and narrow platform about 8 meters higher than the city ground. Restaurants and the ferry terminal are on the ground floor, beneath which the Bund Tunnel is located. The most famous scenic spots along the avenue are the Monument to the People's Heroes at the north end, the Bund Meteorological Observatory in the middle and the Cool Docks at the south end.

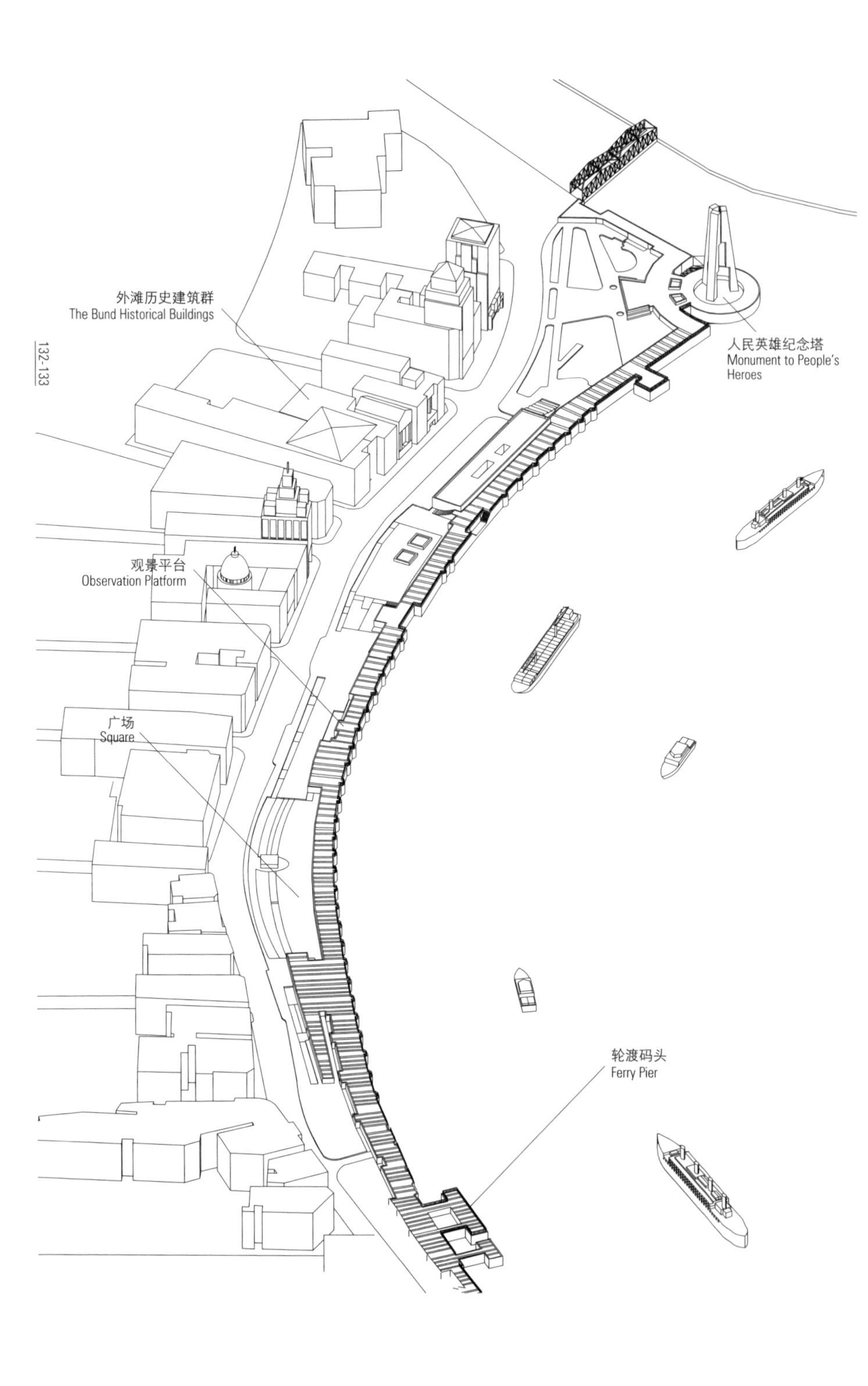
外滩历史建筑群
The Bund Historical Buildings
人民英雄纪念塔
Monument to People's Heroes
观景平台
Observation Platform
广场
Square
轮渡码头
Ferry Pier

作为上海最重要的城市名片，外滩收录了黄浦江对面陆家嘴丰富的现代建筑天际线和身后的历史建筑景观。其自身上下多层功能的叠加，也显示出作为城市基础设施的重要地位。本地居民的日常休憩活动，比如晨跑、夜间散步等，与游客们观光、摄影、购物等活动并行不悖。此类活动在这片巨大的景观平台上构建起一种公共的私密性。

As the most recognized name of Shanghai, the Bund has provided a broad view of both the modern skyline in Lujiazui across Huangpu River, and the historical heritages of the concessions, behind itself. The multi-functional layers overlapped in the Bund have indicated the exalted status as city infrastructure. The everyday relaxing activities of local inhabitants, such as morning jogging and evening strolling, run parallel with the sightseeing and photo-taking of tourists. All these activities on this gigantic sightseeing platform construct a kind of public privacy.

场所：杨浦区怀德路 510 号
功能：住宅
Site: No.510 Huaide Road, Yangpu District
Function: residence

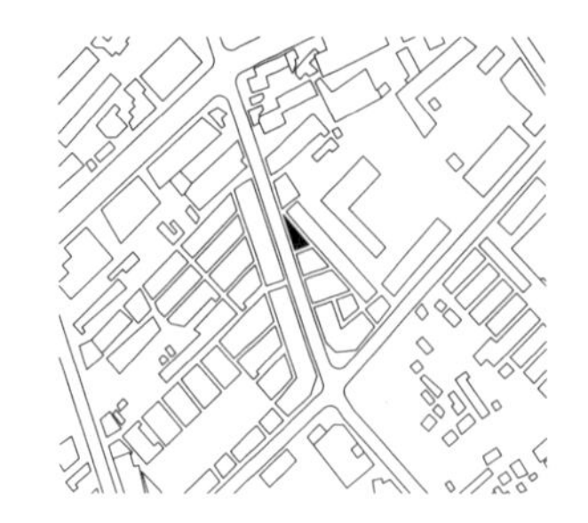

24 三角住宅 Triangular House

三角住宅源自两条呈锐角相交的道路，一条主街，一条内巷。这堵尖墙在朝向主街的一面刷白灰，另一面则裸露出一顺一丁的红砖墙面，上不开窗。住户在屋顶搭建了晒台，而房子朝向主街的一面在首层搭建雨篷形成入户灰空间，可晾晒衣物。

Triangular House was the outcome of two roads, a main street and an inner alley, intersecting in an acute angle. The sharp wall along the main street is lime washed, while the textures of red bricks are exposed on the other side. With limited window openings, the residences added a sunbathing deck on the roof, and awnings on the ground floor along the external side. The awnings have created a transitional space at the entrance, as well as, a place to air clothing.

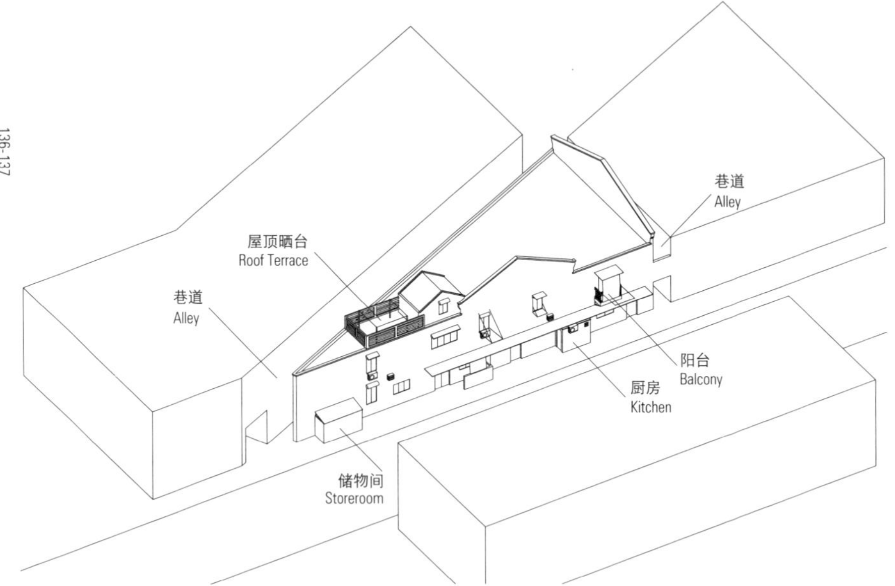

与兵舰头一样，本案例也是由破碎的街块划分导致。在上海老城区还有许多类似的例子，读者可以想象各种它们可能存在的形态。这个三角住宅的户主在建造这堵尖墙以及做各种加建时并非潦草完事，而是考虑了空间对位关系，并精心地修整了各材质的交界处。我们显然能看到，即便是一般使用者在建筑师设计意图之外的改建、增建，同样可以充满设计感。

Much like the Warship-Head-Shaped House, Triangular House is also a result of subdivided blocks that caused by remaining fragments of space. There are many other similar cases in Shanghai's old town, in which one can only imagine the possible shapes of these leftover spaces. The resident of Triangular House did not neglect the details when erecting this sharp wall and making the various additions. The resident has taken into consideration the spatial relationship and the connecting joints of different materials. It is clear that even the alterations or additions of a house by the owner can also present a design feel.

场所：恒通路与共和新路交接处
功能：中学 + 私人钢琴夜校 + 快递配运中心 + 仓储 + 蓄水池
Site: intersection of Hengtong Road and Gonghexin Road
Function: middle school + private piano night school + distribution centre + warehouse + impounding resevoir

25 三明治学校 Sandwich School

这所中学连同它周边的各种设施构成了一个综合体。在中学主入口的背后是位于首层的快递配运中心和仓库。运动场由于建在市政蓄水池上，离地近 2 米高。中学顶层被出租用作独立的私人钢琴学校，它通过一跑独立的楼梯通达首层，并且有独立的入口，与中学在交通上互不干扰。

The complex is made up of the middle school and its facilities, as well as a distribution center and warehouse on the backside of the school. Covering a municipal impounding reservoir is the middle school's playground, which is 2 meter high. The top floor of the middle school is rented out as a private piano night school, which has its own entrance through a separate staircase. This unique layout allows the private piano night school and the middle school to never interfere with one another's traffic flow.

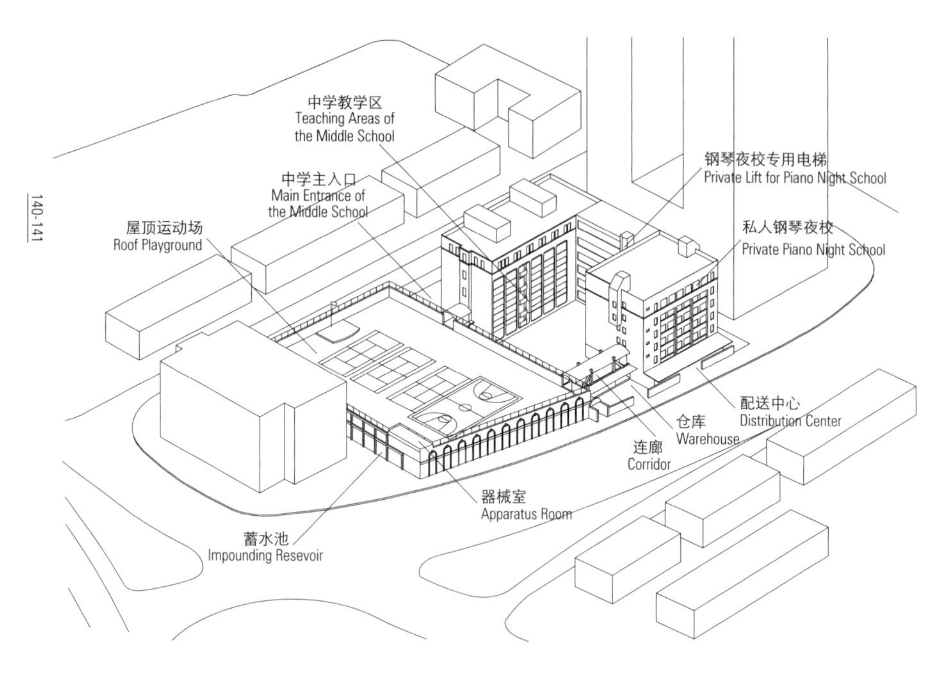

由低到高，蓄水池、快递配运中心与仓库、私人钢琴夜校等功能，像三明治一样层层叠加进这所中学里。城市的基础功能与中学的设施占领不同的标高，相互之间和平共处。中学和快递配运中心在白天运转；而到了晚上，私人钢琴夜校则开始上课。有时候，为了避免干扰四周，货物会在深夜里被卸载和运送进仓库。一种功能在睡眠时，另一种功能往往异常活跃。这就是此类共生功能所谓的“时空错位”。

The name of this project, Sandwich School, refers to the idea that the middle school is overlapped like a sandwich by different functions. From bottom up, the site is organized with an impounding reservoir, a distribution center, a warehouse, and a private piano night school. Moreover, the staggering technique is also used in the program operations, that is, the middle school and the distribution center are running during day time, while in the private piano school gives its lessons in the evening. Sometimes, to avoid disturbing neighbors, goods are downloaded and transported into the warehouse at mid-night. In other words, one function is at sleep while the other is energetic. This method of organization is called "space-time stagger" .

CRE 中铁快运

教学大楼

场所：杨浦区大连路隧道旁
功能：居住（已拆除）
Site: beside Dalian Road Tunnel, Yangpu District
Function: residence (now dismantled)

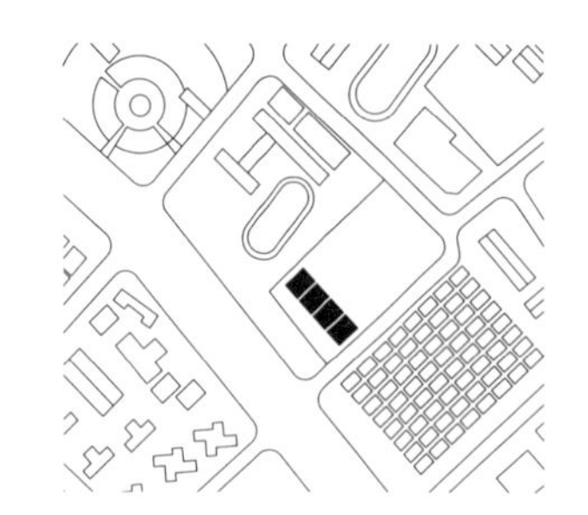

26 四连宅 Four Neighboring Villas

四连宅位于大连路东侧一片拆迁场地上，周边房屋都已被拆，仅存这四间。它们始建于新中国成立前，抹灰外墙面维护得还很好。旧时的街坊们已搬走，但有临时居民搬进来，他们是负责拆迁的建筑工人以及家属，甚至还有一些流浪汉。这里的公共基础设施供给已被切断，临时居民们采取就地取材的方式，充分利用一切触手可及的资源，最终在这片死寂的场地上营建成一个临时的家。

Four neighboring villas are located in an isolated field at the east side of Dalian Road. The villas were erected before the founding of People's Republic of China, in which the original residents have moved away. Temporary residents were workers and their relatives, who were hired to tear down nearby houses. These residents had kept the external walls of the villas in a good condition. However, the supply of public facilities had been cut off, so the temporary residents had to make the most of any resource available in order to set up home in a deserted field, resulting in the condition of the villas to subside.

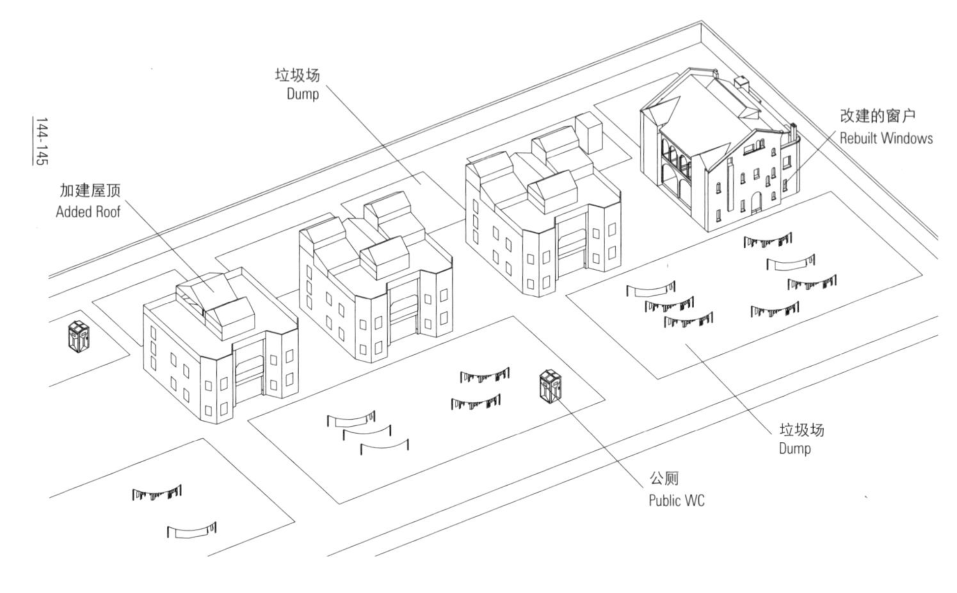

遗憾的是，我们在本书出版前回访本案例，却意外地发现这四栋房子已全被拆除。拾破烂的人在现场留下来的四堆废墟中翻找着尚有价值的物件，破木板与钢筋都被拿去倒卖。四连宅被城市吞噬，底层人群逐渐分解了它们的遗骸。

Upon returning to these four villas, before the publication of this book, we were astonished to find that they had all been dismantled. Scavengers had already came through the four piles of ruins to search for something still of a value, moreover, the broken planks and rebar were collected for reselling. Four Neighboring Villas were murdered by the city, with their bodies being consumed by social desires.

场所：普陀区常德路 1341 号
功能：会所（已废弃）+居住
Site: No.1341 Changde Road, Putuo District
Function: club (now abandoned) + residence

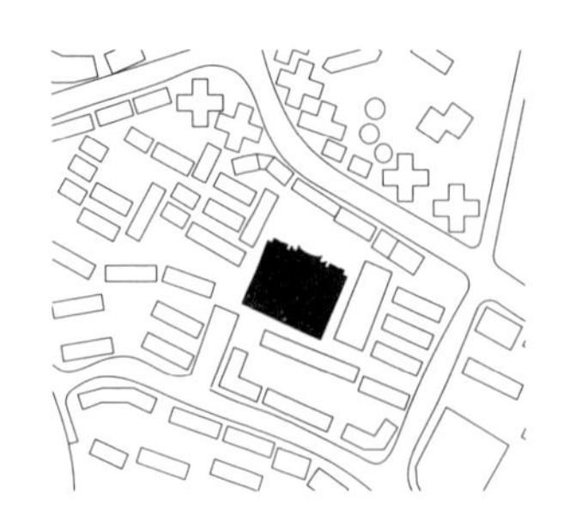

27 忘忧宫
Palace Sans-souci

这栋房子最初是作为会所来设计的，但尚未竣工便遭弃用。它的主立面以及底层一圈都在钢筋混凝土结构外镶贴了模拟西方古典建筑的石材板，入口门廊采用仿科林斯柱式，并且在顶部勾勒了檐口及线脚。当初施工时的脚手架依然没有拆除，各种设备房间在屋顶平台上裸露出来。

This building was originally designed as a club; however, it was abandoned before completion. Stone slates, which are stuck to surfaces of the reinforced concrete structure of the main façade and the ground floor, imitate western classical architecture. It adopts pseudo-Corinthian Orders on the porch, and contains cornices and architraves on the top. Due to the sudden abandonment, the scaffolds still remain and all the equipment rooms are exposed on the roof terrace.

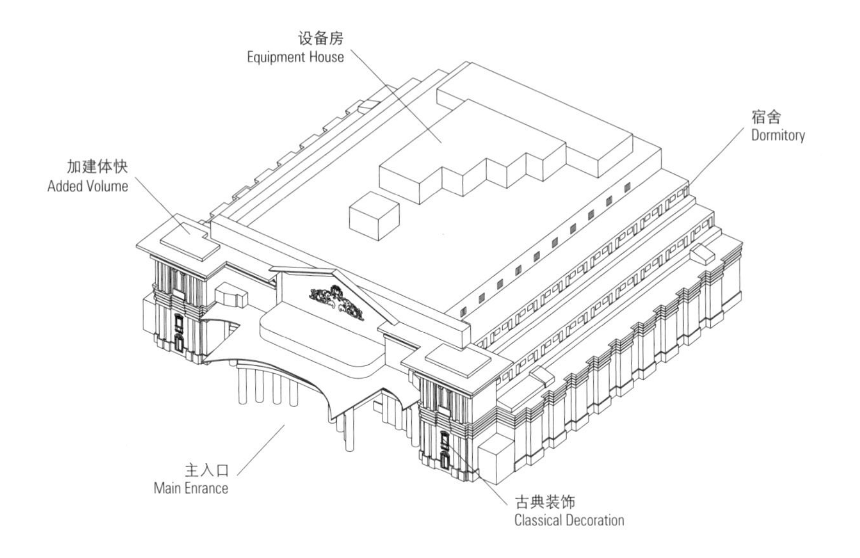

忘忧宫如同一只匍匐于地的大龟。业主挪用了“忘忧宫”这样的欧洲古迹名称，目的显然在于吸引消费人群。为了名副其实，整栋建筑底层贴满了各种古典建筑的符号。然而，它“背上”裸露出来的各种杂乱的发电机、空调等设备以及临时棚屋，都表明业主只在做表面文章。大龟的外壳是消费主义赤裸裸的广告牌，挪用的古迹名称以及仿造的古典建筑风格样式让这栋房屋在成为“山寨”之余，却又反向呈现出消费主义的媚俗原创力。

Palace Sans-souci looks like a gigantic tortoise crawling on the ground. The owner stole the name "Palace Sans-souci" from the original European historical heritage, in which he was aiming at attracting more consumers. In order to make the building "authentic", he adopts all sorts of symbols of classical architecture on the ground floor. However, its presentation is distracted by all the messy facilities, such as electric generators and air conditioners, and temporary shanties built on the back of the palace. The shell of this tortoise has become a bare billboard of consumerism. The stolen name "Palace Sans-souci" and the counterfeit classical architectural style makes this building a cheap copy that presents a kind of kitsch creativity of consumerism.

场所：虹口区吴淞路 560 号
功能：消防站 + 居住
Site: No.560 Wusong Road, Hongkou District
Function: fire station + residence

28 消防站上的棚屋
Shanties on the Fire Station

本案例为虹口救火队原址，现为虹口消防中队营区。该建筑始建于 1867 年，1915 年改建为今日占地 2 660 平方米、总建筑面积 4 400 平方米的规模。建筑呈弧形向内凹进，以便在路口转角留出空地，方便机车进出。建筑为砖木结构，外墙底层和阳台采用石材贴板，其余外立面采用清水红砖贴面。弧面中心设供消防车进出的 4 座大门，在建筑最内侧是 1888 年设立的消防瞭望塔，高 36 米，瞭望半径可达 5 公里。

This building was originally used by Hongkou Fire Brigade. The architecture was at first erected in 1867, and modified to today's scale in 1915. It has an area of 2,660 square meters, and a total floor area of 4,400 square meters. It adopts masonry and wood structure, and its concave-cambered layout leaves a vacant space on the corner of the streets, in order to let the fire engines pass through. Four portals are set at the middle of the arc, providing a passage for fire engines. There is a 36-meter-high watch-out tower at the inner side, which was erected in 1888. On the top of it, one can achieve good eyesight as far as 5 kilometers away.

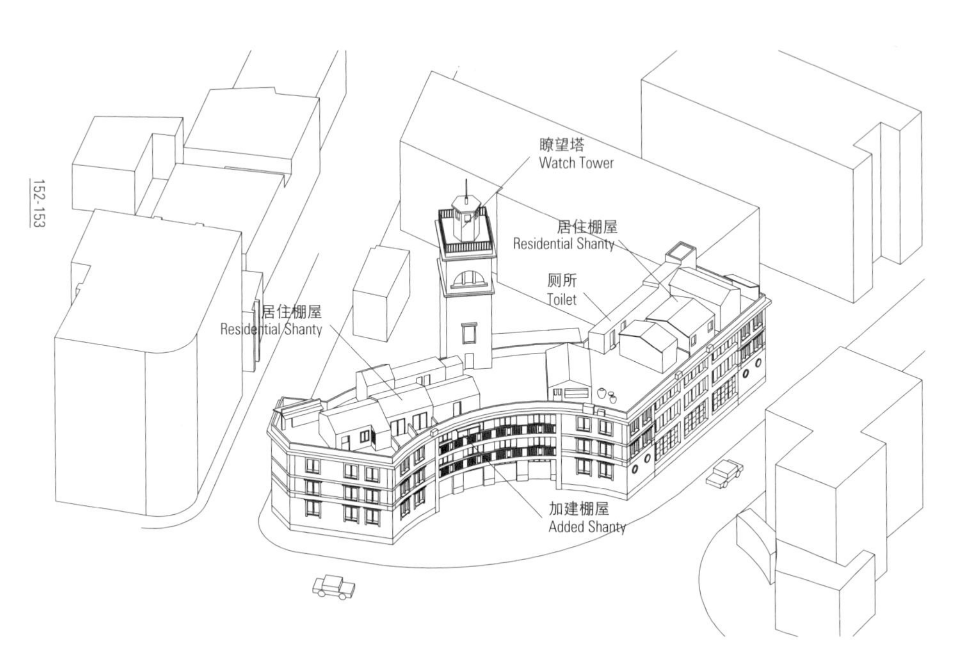

或许对于平日里习惯爬梯的消防武警战士们而言，爬至屋顶是易如反掌的事情。该建筑在屋顶上加建出一些低矮简易的棚屋，以满足若干生活功能（如睡眠、做饭和晾晒）的需求，但基本上没有对这座超过百年的优秀历史建筑造成损害。堂皇与寒酸共存一体，却相处和谐。顶上的棚屋作为消防站的附属功能而合理存在，生活元素不断汇入到历史遗迹中去，缝合了一栋历史建筑与一堆临时棚屋的间距。

Small and crude shanties are added on the roof to meet the demand of several functions relating to living (such as sleeping, cooking and drying); nevertheless, this historical architecture with a history of more than a hundred years has not been damaged. The contrast of the magnificent historical architecture and the shabby addition coexist in harmony with one another. The shanties on the roof, as a reasonable appendix of the fire station, have embedded living elements into a historical heritage, thus reducing the distance between a historical architecture and a pile of temporary shelters.

场所：虹口区沙泾路 10 号
功能：艺术创意产业园 + 商铺 + 餐饮 + 娱乐设施
Site: No.10 Shajing Road, Hongkou District
Function: arts and creative park + shop + restaurant + entertainment

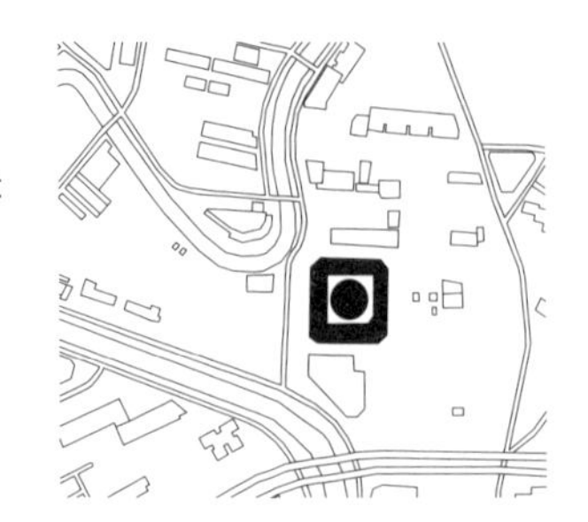

29 上海一九叁三 Shanghai 1933

上海一九叁三老场坊，是 1933 年工部局出资邀请英国建筑师巴尔弗尔设计的宰牲场。建筑外方内圆，东南西北 4 栋建筑围成的四方形厂区与中间一座 24 边形的主楼通过楼梯相连。墙体厚约 50 厘米，两层墙壁采用中空形式，全部采用英国进口的混凝土建造，加工车间采用钢筋混凝土无梁楼盖结构。

Shanghai 1933 was originally the "First Slaughter House in the Far East", whose construction was funded by the Municipal Council in 1933. Its designer was the English architect Balfours. It has a square external shape, as well as, a round inner building. Four side buildings are connected to the main central building by staircases. It is constructed by using the imported cement from Great Britain. The walls are 50 centimeters thick and the inside is hollow. In the workshops, a structure of reinforced concrete flat slab is adopted.

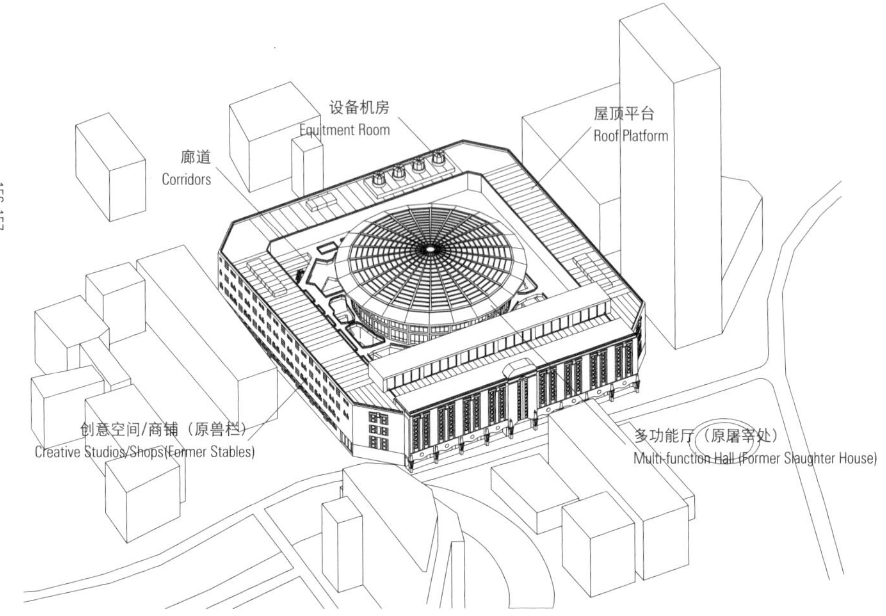

神秘幽深的光线和错综复杂的走道交织成的独特空间，讲述着这座曾被称为“远东第一屠宰场”的建筑的过往。当代人以保护为原则更新建筑——钢和玻璃等现代建筑语汇把它装点成了一座休闲娱乐功能的城市综合体。在极具视觉冲击力的伞形混凝土柱和极尽装饰的精美立面细部中，这栋建筑里的斑驳历史印记全都被保留下来。

Mysterious light and intricate space are created by intertwining passageways that narrate the past stories of this former slaughter house. It was renovated according to the principle of conservation, and modern architectural vocabularies, such as steel and glass, have turned it into an urban complex combining leisure and entertainment. The mottled historical traces in the architecture are kept in both the umbrella-shaped concrete columns, which have a great visual impact, and the exquisite façade details, that provide many decorations.

场所：黄浦区延安东路 260 号
功能：展览＋交通
Site: No.260 East Yan'an Road, Huangpu District
Function: exhibition + transportation

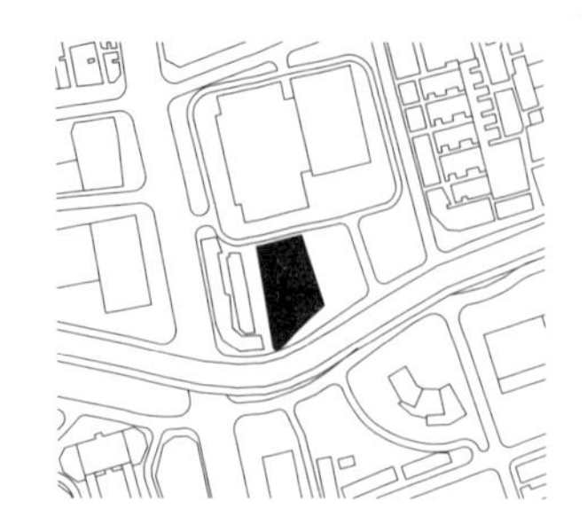

30 高架旁的博物馆 Museum beside an Elevated Road

上海自然博物馆位于延安东路与河南中路的交叉口。延安高架蜿蜒与其擦肩而过，最窄处只相距十几厘米。高架路旁架起高高的隔音挡板，隔离车流的嘈杂噪音。

Shanghai Natural History Museum is located at the intersection of East Yan'an Road and Middle Henan Road. Yan'an Elevated Road meanders beside it with a narrowest distance of mere several decimeters. Sound insulation boards are set up on both sides of the elevated road.

河南中路
右转车辆

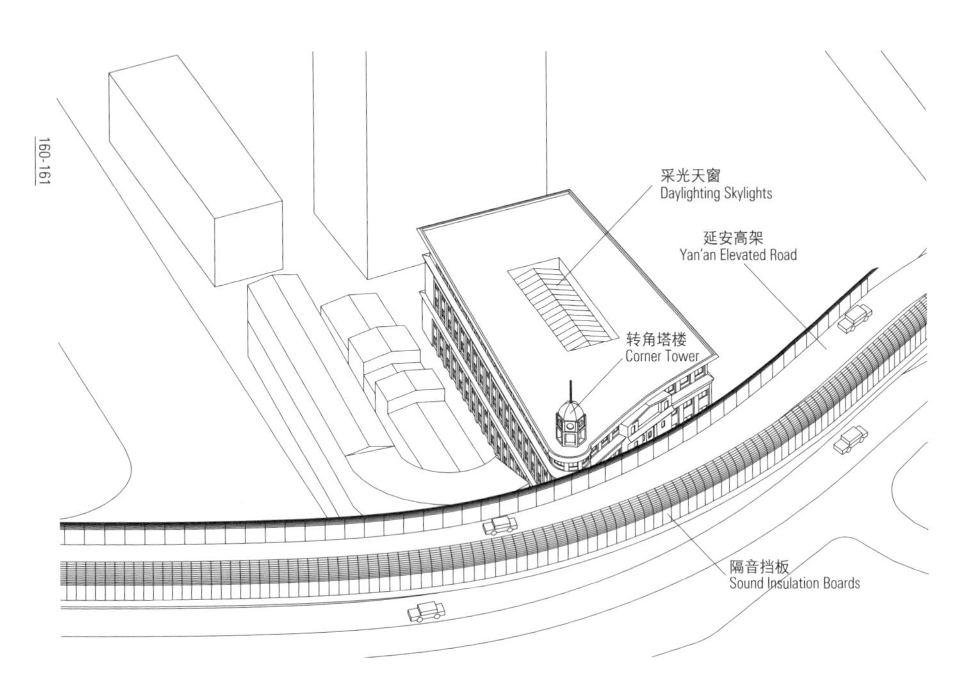

先有自然博物馆，后修起延安高架。它们之间的亲密接触反映出上海城市建设中道路与建筑所能形成的各种极端空间关系。川流不息的车流，呼啸穿过历史建筑的记忆之梦，并置起现实与过往。

Shanghai Natural Museum came into being before Yan'an Elevated Road. The intimate relationship of them two reflects the extreme spatial conditions in the urban development of Shanghai. Flows of cars roar beside the historical architecture, cutting through its dreams, and juxtaposing the cruel reality and pensive past.

WESTIN

场所：黄浦区中华路 581 号
功能：消防预警（已废弃）
Site: No.581 Zhonghua Road, Huangpu District
Function: fire alarming (now abandoned)

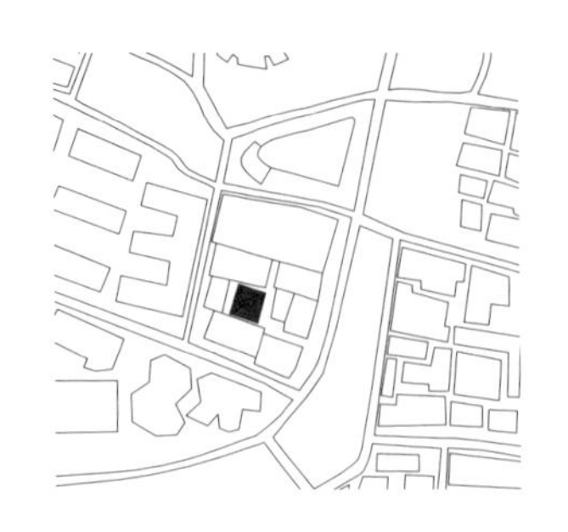

31 消防警钟楼
Fire Station Bell Tower

这座钢筋混凝土结构的警钟塔位于上海老城厢小南门附近，由当时的救火联合队建造于 1910 年。它高 35.2 米，分为 6 层，共设三个小平台与一个大平台，第四层悬挂警钟。各层通过一个由铁笼子包裹起来的螺旋楼梯联系起来，顶层的四向立面密封，只留下高窗。如今警钟楼四周全是居民楼。

The Fire Station Bell tower is located near the Minor South Gate of ancient Shanghai, constructed in reinforce concrete structure by the Fire Brigade Union in 1910. It is 35.2 meters high and divided into 6 floors. There are 3 small platforms and 1 large platform, where the alarm bell is placed on the fourth floor. All the floors are linked by a spiral staircase wrapped in a tall steel cage. The top floor is closed on four sides, only with high windows on the elevation. Nowadays this abandoned bell tower is surrounded by residential buildings.

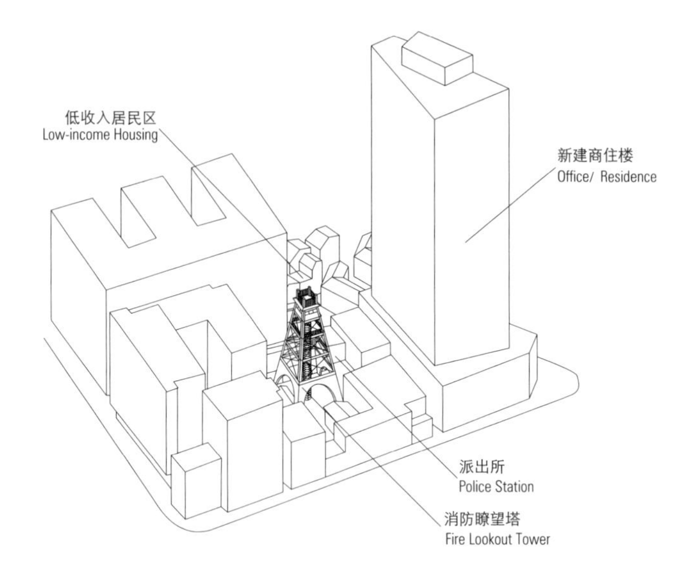

这座消防警钟楼浑身穿戴着混凝土盔甲，曾一度保有该区最高建筑的荣誉。如今虽在建筑高度上风光不再，但它的工业造型依旧使其成为视觉焦点。原有的消防警钟功能早已废弃，现在它与居民楼共生在一起。

Which once was the highest building in this area, Fire Station Bell Tower seems like a giant in a concrete helmet and armor. Even though it was later surpassed by high-rises, it was still a visual focus because of its industrial shape. Its original function of fire alarming has already been abandoned, and now it coexists with the surrounding local residence.

场所：南北高架与苏州河南岸交接处
功能：交通警察局 + 停车 + 修车 + 洗车 + 仓库 + 垃圾处理
Site: intersection of South-north Elevated Road and south bank of Suzhou Creek
Function: traffic police station + parking + car repairing + car wash + warehouse+ garbage disposal

32 高架综合体 II
The Complex under an Elevated Road II

本案例与高架综合体 I 隔苏州河对望。交通警察局位于高架底层空间的端头，前有以栅栏围合起来的一片内院，用来存放违章扣押或遭遇车祸的车辆。在隔壁有一处停车库及修车点，再往前有一处洗车店，而在洗车店的另一侧有垃圾处理场。垃圾车来来往往，扬起的尘土使这一带显得格外脏乱。

This case is facing the Complex under an Elevated Road I across Suzhou Creek. At the end of the ground floor stands the traffic police station, in front of which, there is a yard enclosed by fences. There is a garage center and an auto repair shop next door. Moreover, a bit farther away there exists a car wash, at the backside of which stands a garbage disposal site. Garbage trucks keep coming and going, thus making this area very dirty and disordered.

法兰红

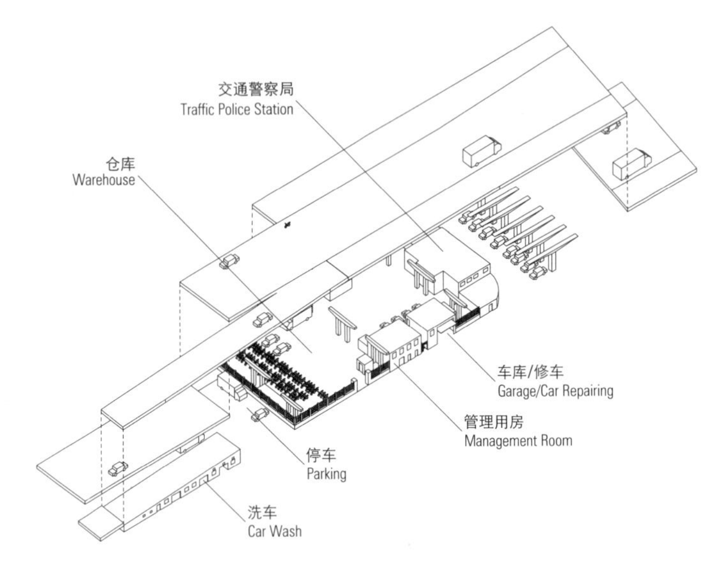

高架综合体 I 中各功能组块偶然地共生在一起，而在本案例中，各功能发生杂交。交通警察局、停车场、修车处、洗车店被统合到高架路底下，构成一个与车有关的“产业”链条。赶来这里的人们可能是到交警局处理车辆事务，顺带可以修理或清洗他们的车辆。

If we regard the functions in the Complex under an Elevated Road I as randomly coexisting, in this case, all the functions are hybridizing. Traffic police station, parking lots, car repairing and car wash are all integrated under the elevated road, thus creating an industrial chain relating to cars. The people who come here may deal with the affairs of their cars, while at the same time they can take their cars to be washed or repaired.

机动车涉牌涉证类交通
交通警察 TRAFFIC POLICE
公安

场所：虹口区塘沽路、大名路、南浔路、长治路围合起的街区
功能：居住＋商业＋旅店＋诊所
Site: block surrounded by Tanggu Road, Daming Road, Nanxun and Changzhi Road, Hongkou District
Function: residence + commerce + hostel + clinic

33 桶中楼 Bucket Housing

桶中楼的外层是三段里弄，内层是一栋混用的大建筑。外层朝外一侧有一间国际青年旅舍、一个诊所以及各种城市服务设施，楼顶上有不少居民自发的违章搭建。内层的建筑体主要用作仓库，但增加了居住、零售等功能，甚至还有小作坊。位于两部分之间的环路空间非常狭小，宽度不到两米，但居民却在当中展开各种类型的日常活动。

Bucket housing is divided into two parts, the external part, which includes three Lilong housings, and the internal part, which is a large single building with mixed functions. There is an international youth hostel, a clinic and other city facilities on the external side. The local inhabitants have made several unapproved additions on the top of these Lilong buildings; while the inner part is used mainly as a warehouse, in which functions such as living, retail selling are added. The circulating space between these two parts is narrow, less than two meters. However, all kinds of everyday activities of local inhabitants take place in such a cramped space.

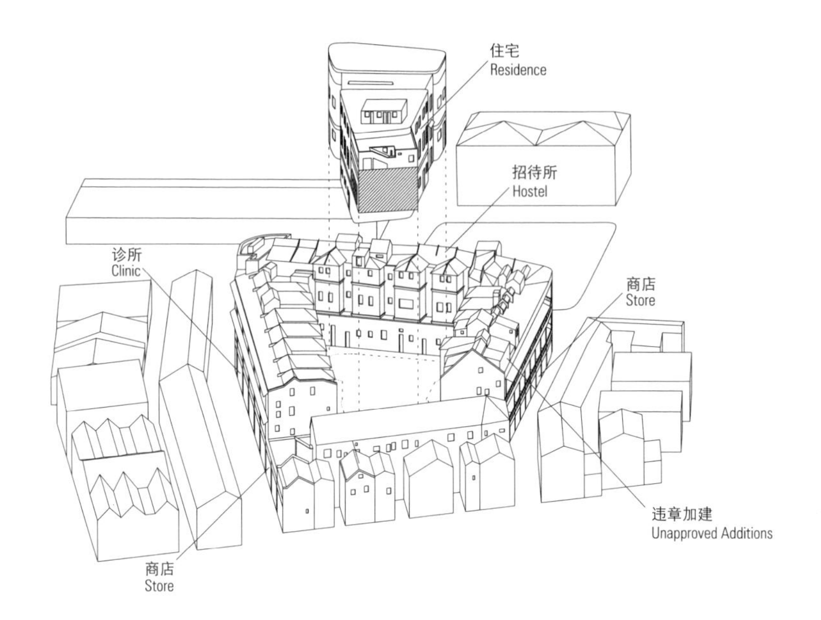

一个满填的街块让地块密度达到了极致，在这里阳光被遮蔽，风雨也被阻挡。而这些狭小的剩余空间得到了非常充分的利用，居民自得其乐，这反过来也说明了在现行的城市法规之外依然可以存在怡人的社区。当每处剩余空间都被居民充分利用起来，这栋楼房就超越了物理意义上的实体空间，而成为一个高密度下的“行为容器”。正是这一种容器，将使用者的行为契合到物态空间中，激发并扩大主体改造空间的能动性。

A filled block makes the density reach a climax. Here sunshine is blocked, and rain and wind are kept away. However, the narrow in-between space has been taken full advantage of. The local inhabitants enjoy their lives here, and their living condition provides a solid proof that there still exists a pleasant community beyond the current city planning laws and regulations. When each inch of in-between space is used by local inhabitants, the building surpasses a mere physical space and becomes a high-density "container". It is this container that has embedded the behaviors of users into physical spaces, thus stimulating and extending the initiative in space alteration.

长治路
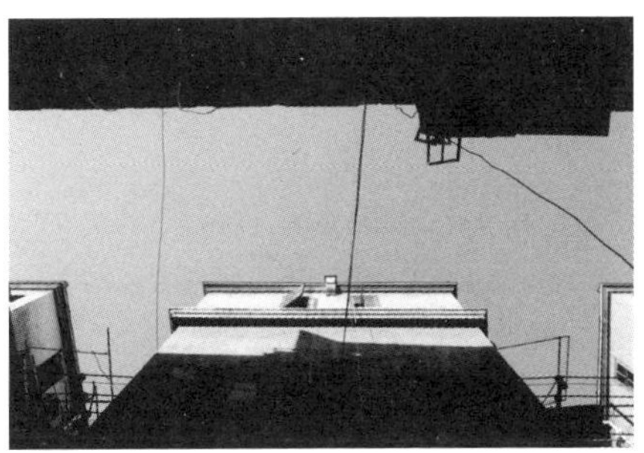

场所：虹口区九龙路 270 号
功能：水闸 + 办公
Site: No.270 Jiulong Road, Hongkou District
Function: sluice + office

34 水闸办公室 Sluice Office

本案例横跨在汇入黄浦江的俞泾浦河道上。这栋房子由三部分组成：位于水面上的中间主体、下面的水闸，以及连接两旁路面上的侧翼。屋顶上高高扬起充满机械感的水闸操作臂，水闸每天开闸两次，一次在中午一点左右，另一次在晚上八点左右。中间的房子是工作人员的办公室，他们在房子周围栽种各种植物，并且饲养一些鸟类，将周边环境布置得很惬意。

Sluice Office is located right above the channel of Yujing Creek, which flows into Huangpu River. There are three parts in this case: the main building above the water in the middle, the sluice below it, and the lateral wings connecting to the roads on both sides. High on the roof stands the mechanical manipulator arm of the sluice. The sluice opens twice a day, around one o'clock in the afternoon and around eight o'clock in the evening. The main building is used as the office for the sluice staff. They grow various plants around the building and also raise some birds. The environment has been made very pleasant.

国联通
SIPG
外滩隧道
Bund Tunnel
中山南路方向
延安东路方向

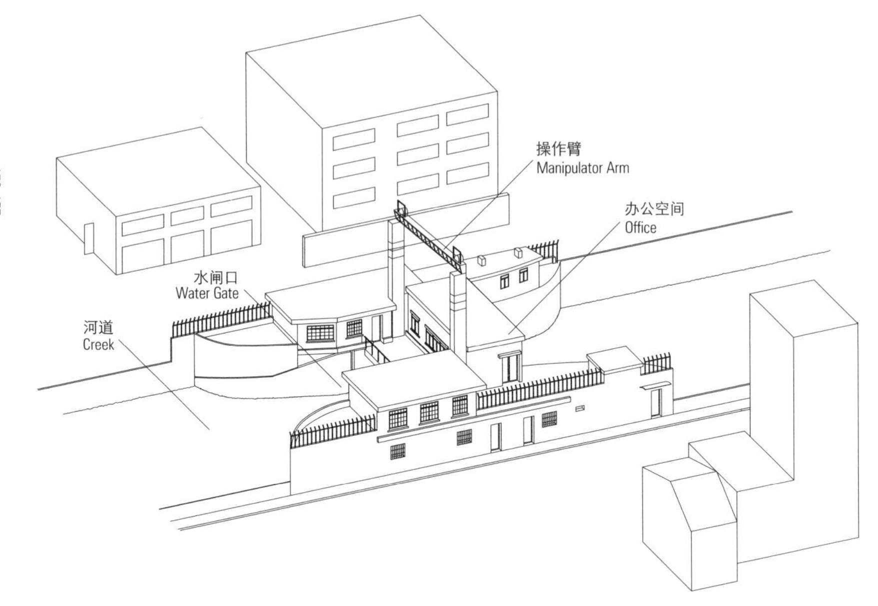

本案例将河道基础设施、工作人员的操作空间以及与两侧地坪的交接面整合成一个连续体，极大地改造了原本由机械设施带来的消极空间。它最大化地利用了剩余空间，将人的生活模式注入原本冰冷的物质中，为空间带来生气。使用者在超出设计者之外的自发改造行为必然导致某种不确定，但最终呈现出来的空间配置却又是如此的合情合理。

This case has integrated the channel infrastructure, operational space of the staff, and the connecting part with the bilateral ground, resulting in improving the negative space caused by the mechanical facilities. It has optimized and revived these in-between spaces by injecting the warm living mode of human beings into the cold empty spaces created by infrastructure. It is obvious that such spontaneous alterations by the users will lead to uncertainty, while the result of the spatial distribution seems quite reasonable.

外滩隧道
Bund Tunnel
中山南路方向
延安东路方向

场所：杨浦区国顺东路 800 号
功能：艺术工坊 + 旅业 + 酒吧 + 茶室 + 洗车
Site: No.800 East Guoshun Road, Yangpu District
Function: art studio + hostel + bar + teahouse + car wash

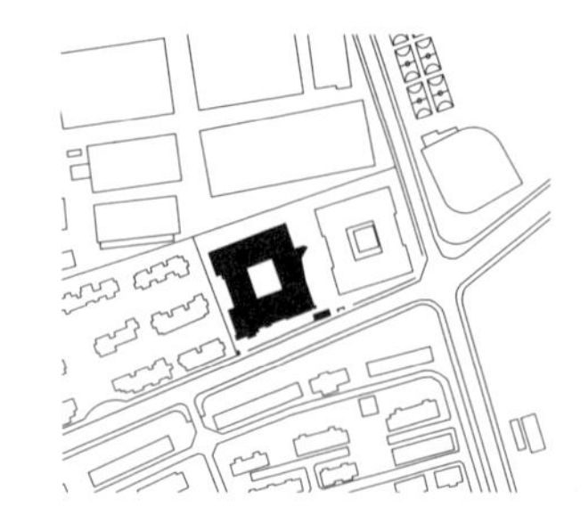

35 艺术方盒 Art Box

本案例原名“上海艺术区 800”，因总体造型为方盒而得此昵称。园区主大楼共 6 层，总建筑面积约 21 460 平方米。一楼层高 5.5 米，二至五楼层高 4.5 米，六楼层高 6.9 米。一楼中庭设置可容纳 180 人的多功能展演空间。其屋顶与三楼地面齐平，形成面积达 365 平方米的开阔室内平台，上有可遮光式的采光罩，内有冷暖气，全年皆可使用，适合举办大型开幕酒会与艺术展览。

The real name of this case is "Shanghai Art Zone 800", and its nickname "Art Box" comes from its cubic shape. The main building is 6-story high, with a total floor area of 21,460 square meters. The central grand hall on the ground floor is an exhibition space capable of accommodating 180 people. Its roof is at the same height of the third floor, where lies a wide indoor terrace with an area of 365 square meters. With an overhead skylight, central air-conditioning and heating, this terrace is perfect for holding large-scale opening cocktail parties and art exhibitions.

中和堂
P
T:55081993

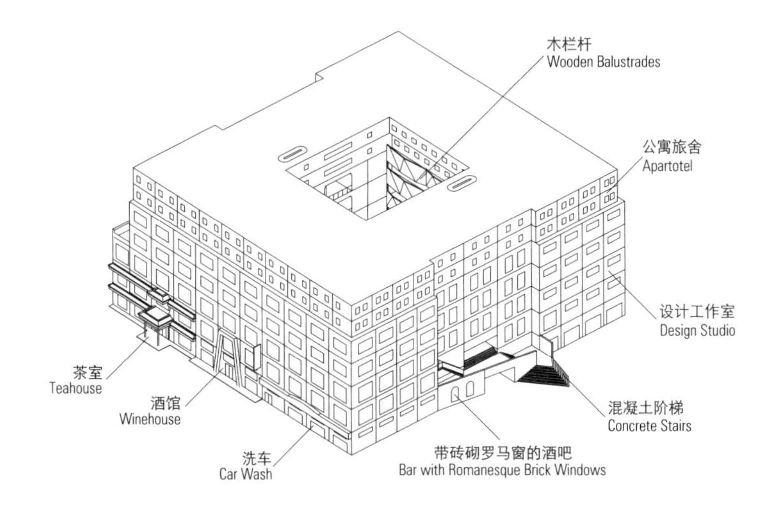

艺术方盒在闹市中呈现出一个宁静且与世隔绝的艺术世界。底层的酒吧、茶室等多种休闲设施为在上面工作的艺术家们提供了放松与交流的空间，此外，在房子的东北角有一家公寓旅舍。毫无疑问，这个汇集以上功能的密实立方体成为艺术家们的世外桃源，因为一切世俗的纷乱嘈杂都被挡在门外。一旦跨入这个艺术方盒，就能感觉到被艺术的氛围所包裹。

Art Box presents a tranquil and isolated world of art within this lively urban environment. Various facilities such as a bar and a teahouse are placed on the ground floor, providing a relaxing and communication space for the artists working above. Furthermore, a hostel is located at the north-east corner of the building. All these functions within a single compact cube, without a doubt become an arcadia for artists since all the profane noises and turbulence outside have been completely kept out. Once one gets into this art box, he will feel himself embraced by a religion of art.

五角场800号
东楼

场所：闸北区成都北路 1085 号
功能：办公
Site: No.1085 North Chengdu Road, Zhabei District
Function: office

36 水塔楼 Water Drainage Building

水塔楼位于苏州河右岸，被周边的房屋以及旁边穿过的南北高架掩盖。它是一座独栋房子，尽管体量小，但圆柱状的外表很引人注目。这栋房子夹在办公楼以及工业厂房之间，上层有条形窗以及露台，下层有大门与大烟囱。岁月带来的微妙变化使之更为有趣。它那原本纯净的几何外形因各种需要而改动，比如加出不同屋顶、增加机械设备、外楼梯处的蓝色体量，等等。

Water Drainage Building is located at the right side of Suzhou Creek. Despite its small volume, this independent building is conspicuous for its cylindrical shape. Between an office building and an industrial plant, Water Drainage Building has ribbon windows and terraces on the upper floor, with a large portal and chimney on the lower floor. Likewise, the subtle changes brought by time make it more charming. The original pure geometric volume has been changed for different needs; such as, the roof additions, the newly installed machines, and a small blue volume outside the external staircase.

上海排水公司
南苏州路

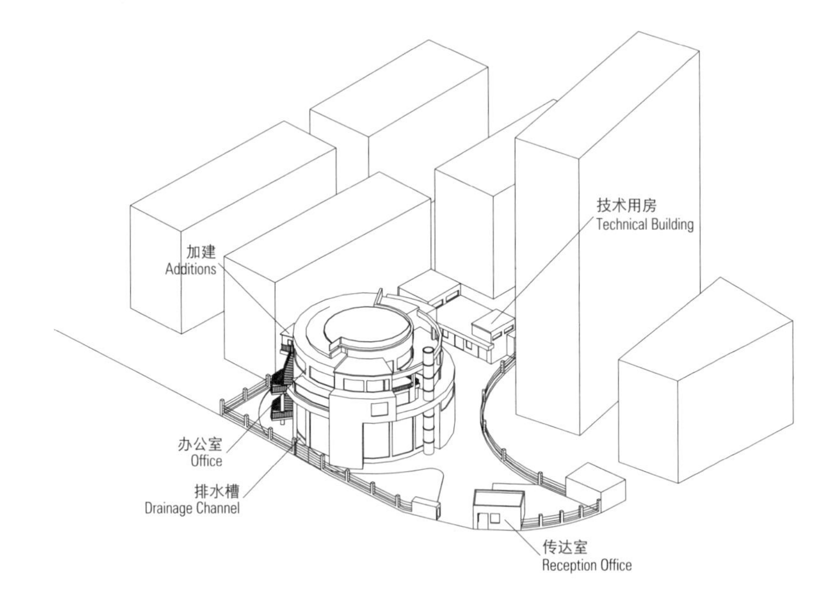

水塔楼主要供上海自来水公司的员工办公，满布其上的加建反映出使用者对空间利用的自我主张。这些加建出来的体量主要供储物之用，它们与原有的柱状外立面也形成良好的比例关系。圆柱状体量是上海自来水公司办公楼很常见的造型，而后来的加建让它更加丰满。

Water Drainage Building is mainly used as an office for the staff of Shanghai Water Supply Company. The additions reflect the self-assertion of the users in their own uses and needs of the space. Most of these newly added volumes are for storage, and have formed a good proportion with the original cylindrical façade. Generally speaking, a cylindrical volume is often utilized in the office buildings of Shanghai Water Supply Company, while the later alterations and additions make it diverse.

上海排水公司
工程抢险

场所：虹口区虹镇老街
功能：居住
Site: Old Hongzhen Street, Hongkou District
Function: residence

37 虹镇老街 Old Hongzhen Street

虹镇老街曾是上海中心城区最大的棚户区，现在是政府重点拆改的地段。这栋房子是虹镇老街很典型的一个实例，底层是服装店，上层用作居住。各种加建比比皆是，陡峭的钢制楼梯通往顶层，窗洞上是刚更换过的铝合金窗框。窗台上的绿色植物，散发出浓浓的生活气息。

Old Hongzhen Street was once the largest shanty town in central Shanghai, and now it has become an area for renovation. This house is a typical case for this street. Its ground floor is used as a clothing store, with the upper floors for living. There are various later additions, such as a steel staircase leading to the top floor and the aluminum sashes added to the original window openings. The plants on the window sill bring vitality into the space.

光酒家
正宗
河南拉面

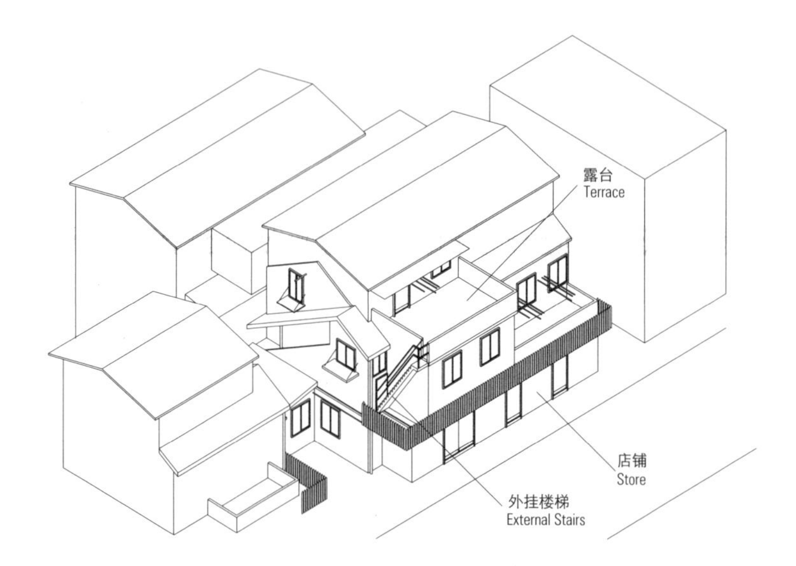

上海近年来的旧城改造运动酝酿了一个个街区死与生的故事。住在暗无天日、卫生条件堪忧的棚户里的居民有机会搬迁到新居开展新生活，无疑是一件好事。但操之过急的拆迁工作往往没有对街区里的建筑做好足够的评估，导致很多好房子没能保留下来。此外，接踵而至的大型楼盘也将抹杀掉多姿多彩的街道空间以及社区活力。

The recent renovation of the old town in Shanghai has made various stories of deaths and lives of streets. There is no doubt that the inhabitants living in these dark and dirty shanties will benefit from having a chance to move out to a new residence. However, the assessments of the buildings are often not enough due to the overhastiness in the renovation, thus leading to the dismantlement of many high-quality buildings. Additionally, the coming large housing properties will wipe out the various street spaces and energy of communities.

复苑教育
初高中小班化辅导专家
官方网址：www.fudanyuan.com
天一服饰

CHIGO

场所：虹口区溧阳路 611 号
功能：焚烧（已废弃）
Site: No.611 Liyang Road, Hongkou District
Function: incineration (now abandoned)

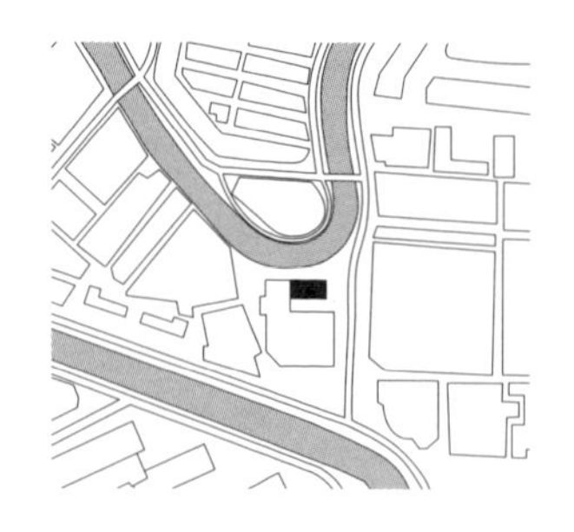

38 焚烧炉 Incinerator

这座焚烧炉在 20 世纪 30 年代与上海一九叁三一同建造。它占地约 600 平方米，呈矩形布局，高 4 层，每层约 3.6 米高。一根高大的烟囱矗立在建筑背面，直临俞泾浦河面。外立面采用与一九叁三类似的混凝土材质，但更多地采用上悬铁窗。竖向线条在建筑外立面上形成有节奏的韵律，混凝土、钢、玻璃这三种主要材料构成一曲空间质感的合唱。

This incinerator was erected together with Shanghai 1933 in the 1930s. It covers a rectangular area of approximately 600 square meters and contains four floors, each of which is 3.6 meters high. A gigantic chimney stands at the back of the building, facing directly towards Yujing Creek. Its façades adopt the cement materials similar to Shanghai 1993, but employ more top-hung iron windows. The vertical lines repeating on the façades have formed a kind of rhythm, and three main materials (concrete, steel and glass) constitute a chorus of spatial textures.

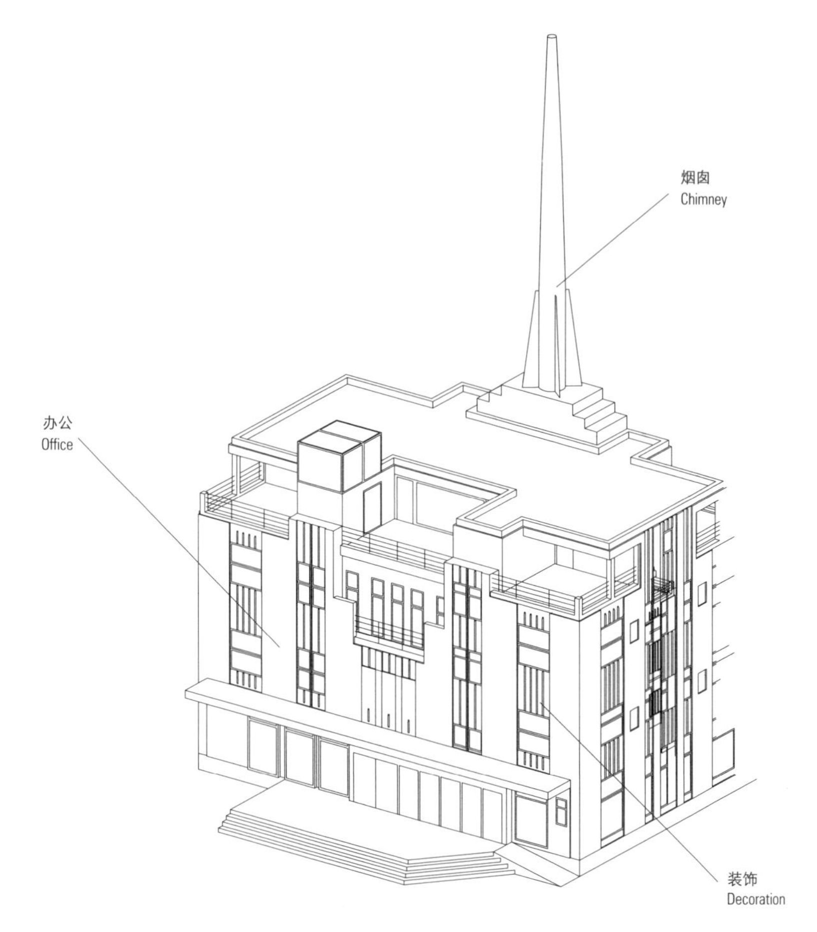

作为屠宰场的配套设施，这座焚烧炉在当时算是巨型的工业建筑。它位于一九叁三的隔壁，时至今日，当两者之间的功能联系已经不复存在后，它们之间就只剩下空间上的对望。与它的邻居命运不同，焚烧炉现在被用作不对公众开放的办公室，导致周边空间很消极。下一轮的城市更新能或许能为它置换进更适合的功能。

As a supporting facility of the slaughter house during that time, this incinerator was a gigantic industrial building. The functional relationship of these two buildings no longer exists today, and only their confrontation in the space remains. Different from its neighbor, this incinerator has now been changed into a private office, causing the surrounding area to be very negative. Perhaps it can be filled in a more suitable function in the next round of urban renovation.

场所：黄浦区中山南路 1157 号
功能：商务酒店（现荒废）
Site: No.1157 South Zhongshan Road, Huangpu District
Function: business hotel (now abandoned)

39 摩登假日
Modern Holiday

摩登假日商务会所位于南浦大桥盘旋高架的北侧，总面积约为7 000 平方米。在朝向南浦大桥一侧的主入口处，它在立面上用钢框与黑玻璃悬挑出一个巨大的“之”字形体块，内部是一个通高的大堂。立面另一侧沿用了这种构图元素，设计了五排锯齿状的窗户。

The business hotel of Modern Holiday is located at the north side of the Spiral Viaduct of Nanpu Major Bridge. It occupies an area of about 7,000 square meters. On the main entrance side, which is facing Nanpu Major Bridge, it has a gigantic cantilevered "Z-shaped" volume made of steel frames and black glass, followed by a lobby of full-height. The same elements are also adopted on the left part of the elevation, thus leading to 5 rows of windows in zigzag shape.

洗浴 客房
摩登假日
摩登假日

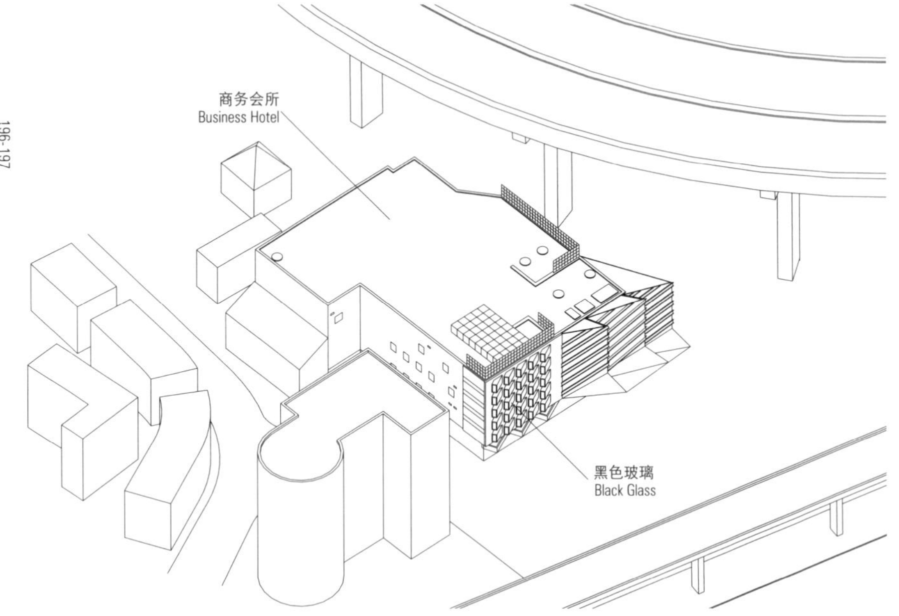

摩登假日给街道提供了一个纯粹性质的立面，一个夸大自身的语言系统：在外面看起来是 6 层的建筑实际只有 3 层。然而，这个曾经辉煌的酒店兼娱乐城现在已经停业快两年了。它与上海其他废弃建筑一样，难以逃离被即将入驻的新业主拆除的厄运。可以看到，消费社会无止境的物欲洪流席卷一切。

Modern holiday presents a purely decorative façade to the street. It seems like an exaggerated lie, seeming to have 6 floors when viewed from outside, but actually there are 3 floors. However, this hotel and entertainment center has been abandoned for almost two years. It is facing a similar fate with other deserted buildings in Shanghai and can hardly escape the doom of being dismantled by the coming new owner. It clearly indicates that the endless raging flood of consumer society is devouring everything.

场所：虹口区赤峰路 325 号
功能：菜场 + 网吧 + 旅馆
Site: No.325 Chifeng Road, Hongkou District
Function: food market + internet bar + hotel

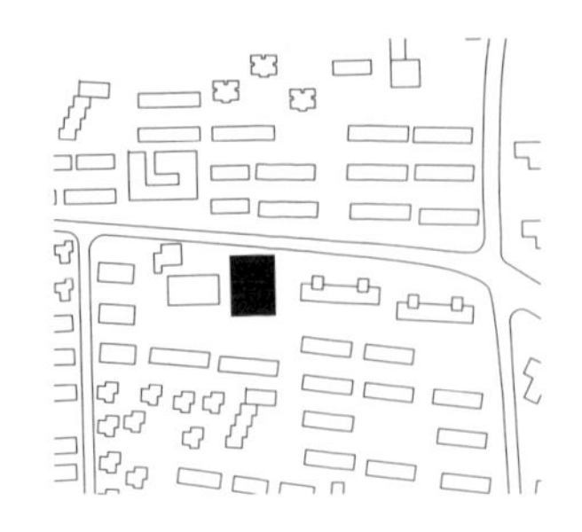

40 菜场旅馆 Market Hotel

上海虹口区的连锁菜场以“三角地”这个百年老字号命名，它们往往会与其他功能混杂起来，赤峰路上的这家最为典型。这栋建筑共 5 层，梁柱结构。菜场占满整个首层，约 800 平方米；二楼是网吧；菜场东侧有一个独立的入口通到位于三到五层楼的旅馆。菜场前总是停靠着各种卸货的面包车或者小贩的三轮车。

The chain food markets in Shanghai Hongkou District are given a time-honored brand name "Triangular Terra". They are often combined with other functions and this example on Chifeng Road is a typical case. This building has five floors, all using a post and beam structure. The food market fills the entire ground floor, covering an area of about 800 square meters. An internet bar is placed on the first floor and a separate entrance on the east of the market leads to a hotel located from the second to fourth floors. Usually in front of the food market are parked various unloading vans and tricycles.

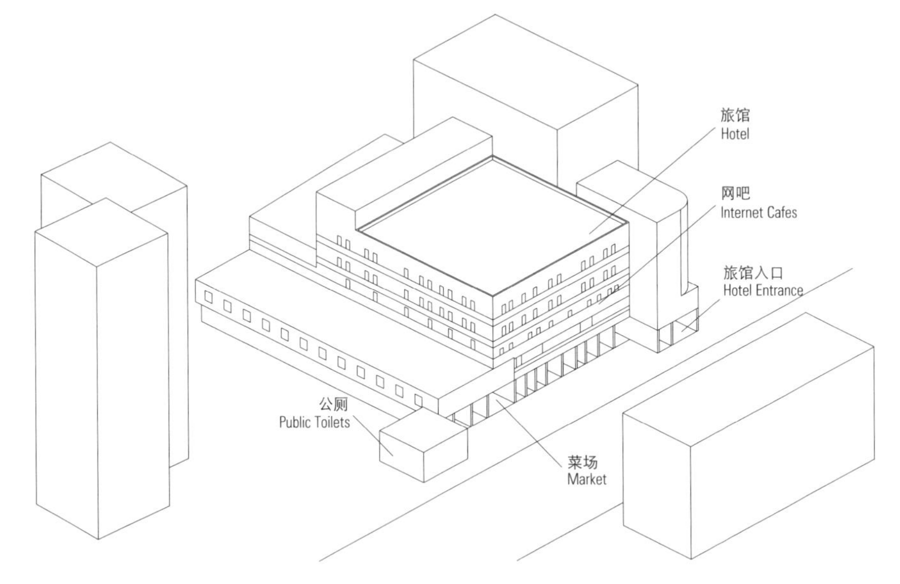

这是一个上海常见的若干种居住配套功能自下而上叠加到一栋大楼中的案例。将本身喧哗的菜场功能与需要宁静的旅馆功能并置，初看似乎很不合理，但它们只要通过“错峰用电”就能实现互不影响。常常一种功能在睡眠的时候，另一种功能异常活跃。白天，菜场人声嘈杂；晚上，菜场关门，旅店获得了安静的环境。

This is a common case in Shanghai in which several living-supporting functions are put together from bottom up inside one single building. At first sight, it seems quite unreasonable to juxtapose a noisy food market and a hotel, which requires a quiet atmosphere. Nevertheless, by adopting an "off-peak" method these functions can avoid interfering with each other. Usually a function is dynamic while another is asleep. The market is crowded and noisy during the daytime, while at night the food market is closed, allowing the hotel above to gain peace and quiet.

上海三角地菜场
预约电话

上海三角地菜场
SHANGHAI SANJIAODI FOOD MARKET
曲阳分场

上海三角地菜场
SHANGHAI SANJIAODI FOOD MARKET
曲阳分场

58
168

功能：虹口区中山北一路 678 号
场所：商铺 + 办公 + 宿舍
Site: No.678 First North Zhongshan Road, Hongkou District
Function: shop + office + dormitory

41 高架小楼
Building beside the Elevated Road

全长 48 公里的上海内环高架路在中山北一路转入北二路这一段上与一栋小楼相距非常近，最近处只相隔 30 厘米。高架小楼底层有商铺，二、三层是办公室，顶层部分出租作为宿舍，居住者大多为外来务工人员。为避让高架路，小楼在三、四楼都有退台处理，因此形成的室外小平台常被用来晾晒衣物。虽然高架路两侧有隔音挡板，但依旧不能完全隔绝噪音。

The Shanghai Inner Ring Elevated Road is in close proximity to a building from First North Zhongshan Road into Second North Zhongshan Road. The shortest distance is a mere 30 centimeters. There are shops on the ground floor of this building, while offices are located on the first and second floors. A part of the third floor is rented out as dormitories with most of the tenants being migrant workers. The building has setbacks on the second and third floor in order to avoid the elevated road, and these small outdoor terraces are often used for airing clothes. Though there are sound insulation boards on both sides of the road, noise still cannot be totally insulated.

院内
限速
5
摩托车
助动车
自行车
请勿入内

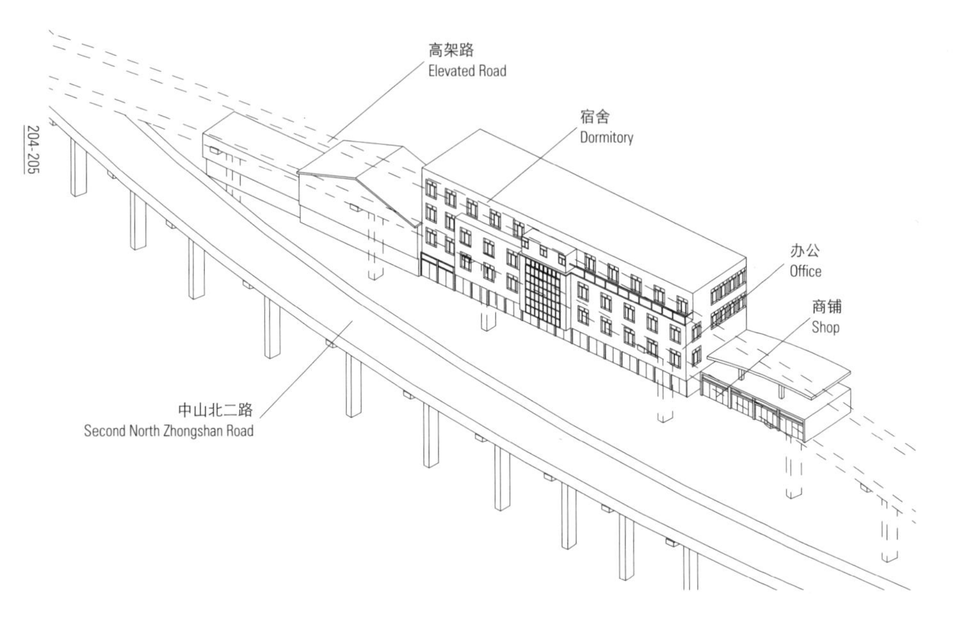

在居住高密度之余也伴随着交通的高密度发展。高架路直接从高空中切过既存的城市街块，但它与底下的社区共存而相安无事。本案例是一个城市基础设施与日常建筑共生的例子。建造在后的小楼毫不惧怕身旁的“高架怪物”，它充分利用了自身的用地红线内的每一寸空间，以致与高架路形成如此直接的对峙。

High density in living space is accompanied with that in transportation. Elevated roads directly cut through the urban blocks in the air without disturbing their neighbors below. This case is a good example of the coexistence of urban infrastructure and everyday architecture. The later-built building shows no fear of the "elevated monster" passing by. It takes full advantage of each inch of space within its property line, thus resulting in such a direct confrontation with the elevated road.

场所：虹口区唐山路 3 号
功能：住宅 + 商业
Site: No.3 Tangshan Road, Hongkou District
Function: residence + commerce

42 街角楼 Corner Block

街角楼是占满一个街区的一整栋楼。它位于五条道路的交叉口，特殊的位置造就了建筑特殊的形态。建筑首层用作商业，共有 19 户商家；二三层为住宅，每层有住户 23 户。原有的平屋顶上布满了自主搭建，共加建 12 户，使得整栋建筑呈现臃肿混杂的状态。

Corner Block is one single building, which occupies an entire block. Located at the intersection of five roads, it has formed a special shape. The ground floor is used for commerce, including 19 shops. There are local inhabitants living on the first and second floor, each of which accommodates 23 families. Later unapproved additions have covered the original flat roof, thus making the whole building look over congested.

好得来
劳防专卖
电话：13817877778
30

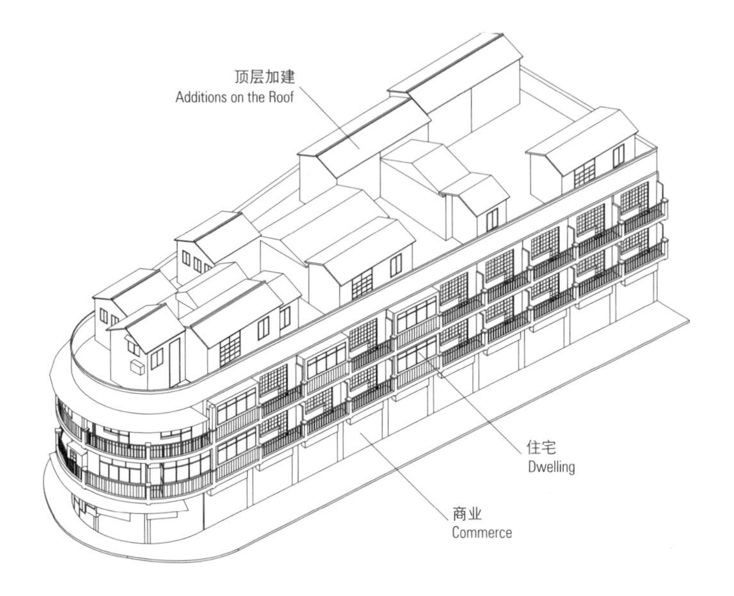

在某种意义上来说，街角楼是中国式的“街道对景”，公共性与私密性交叉在一起。在前端半圆形部分，居民在阳台上的活动完全暴露在城市的公共目光中。倚在栏杆上，居民俯瞰着这个位于五叉路口上的城市街区、街道上来来往往的人流、街对面商店里倾泻出来的水果摊。街道上的行人也会看见阳台上的居民——穿睡衣的女人，弄鸟笼的男人——并窥到他们生活的一角。

In a certain sense, Corner Block is a typical Chinese "Street Scenery", with publicity and privacy crossing together. In the front semi-circular part, the activities of the inhabitants on the balconies are totally exposed to the city. Leaning against the balustrades, the inhabitants can overlook the five-road-crossing urban block below to view the flow of people coming and going in the streets, in addition to the fruit stands in the shops opposite the street. Passers-by in the street catch sight of the inhabitants in the balconies, such as a woman in her pajamas and a man beside his birdcage. A part of their private life is exhibited to the public.

中国移动通

中国移动通

场所：黄浦区花园港路 200 号
功能：展览
Site: No.200 Huayuangang Road, Huangpu District
Function: exhibition

43 艺术发电厂 Power Station of Art

上海当代艺术博物馆（号称“艺术发电厂”）位于黄浦江西岸。建筑主体长 128 米，宽 70 米，高 50 米，原建筑面积为 31 000 平方米。高达 165 米的钢筋混凝土烟囱笔直高耸于东北侧。它前身为始建于 1897 年的清末南市电灯厂，1955 年改名为南市发电厂，1985 年时这个巨大烟囱建成。2010 年上海世博会将它改建成城市未来馆。2012 年后，它已成为运行良好的公共艺术平台。

The Shanghai Contemporary Art Museum (known as "Power Station of Art") is located to the west of Huangpu River. Having an original floor area of 31000 square meters, it is 128 meters long, 70 meters wide and 50 meters high now. The 165-meter-high smokestack made of reinforced concrete stands on the northeast side. Formerly it was the Nanshi Light Bulb factory erected in 1897 at the end of Qing Dynasty. In 1955 it was renamed Nanshi Power Station, and set up this gigantic smokestack in 1985. It was turned into the City Future Pavilion during the Expo 2010 Shanghai, and has been a well-operated public art zone since 2012.

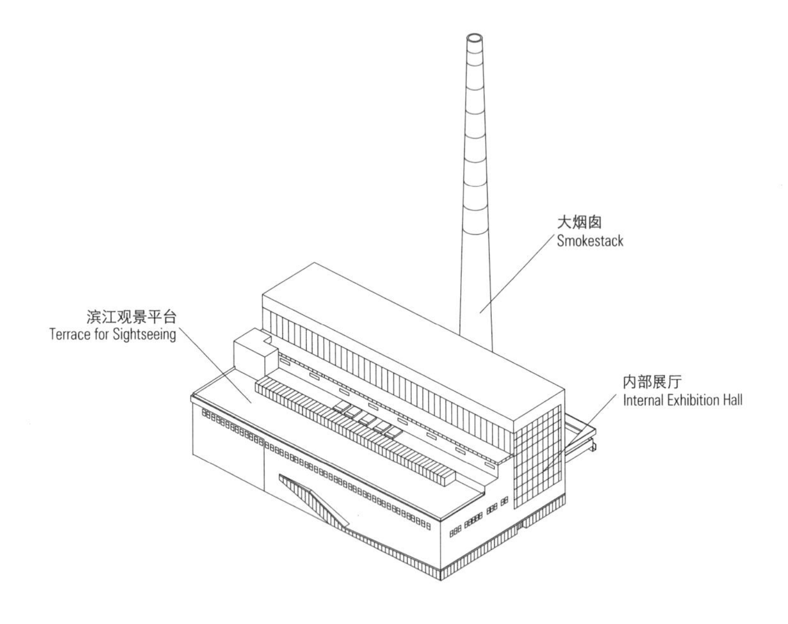

从工业厂房到艺术展馆的功能置换让这栋体量庞大的建筑重焕生机。巨大的室内空间带着工业的力量感，映照在底下的当代艺术展品上。从以前的工业发电厂到世博会期间的展示馆，再到如今的文化“发电”中心，上海当代艺术博物馆见证了整个这一系列特殊事件。

The displacement of the function from an industrial factory to art pavilion has rejuvenated this gigantic building. Its tremendous interior space has brought in a strong sense of industrial power, which is shining upon the contemporary art works below. Shanghai Contemporary Art Museum has witnessed a variety of special events, varying from a former power station, to a pavilion in Expo, and also to the current center of the "generator" of culture.

入口

场所：杨浦区辽源西路 111-133 号
功能：居住
Site: No.111-133 West Liaoyuan Road, Yangpu District
Function: residence

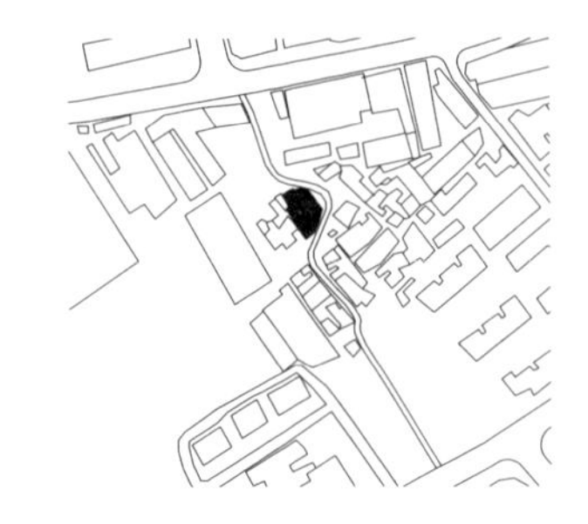

44 弯巷 Curved Alley

“弯巷”是对一个占地不足 300 平方米的高密度棚户区的昵称，它躲藏在一条弯弯曲曲的巷道后面。弯巷的几个面有不同的变化：在面对巷道的东面，实墙居多，仅留高于人视线的小窗，构成一个封闭的巷道立面；在南面则完全敞开，阳台、大窗、不规则且与公共道路无边界的小院子，居民在里面种植物、晾晒衣服。院子的归属很模糊，半共半私，每一个路人都可以进入与居民交谈，而住户也在这展开一些私密的活动，比如吃饭。屋顶面也是弯巷的一大特色，朝向天空的地方被不同高度的屋顶天台占领。

"Curved Alley" is the nickname for a high-density shanty residence of about 300 square meters. It differs on each side. On the east side facing the alley, there are only walls with small windows above sight line, thus making itself a blocked elevation. Yet on the south side, the residence is completely open, containing balconies, large windows, small irregular courtyards of no boundary with the external road, etc. Local inhabitants are growing plants and drying clothes in this location. Such courtyards share an ambiguous ownership of both publicity and privacy. Roof terraces of various heights occupy different areas facing the sky.

中国凉茶
和其正

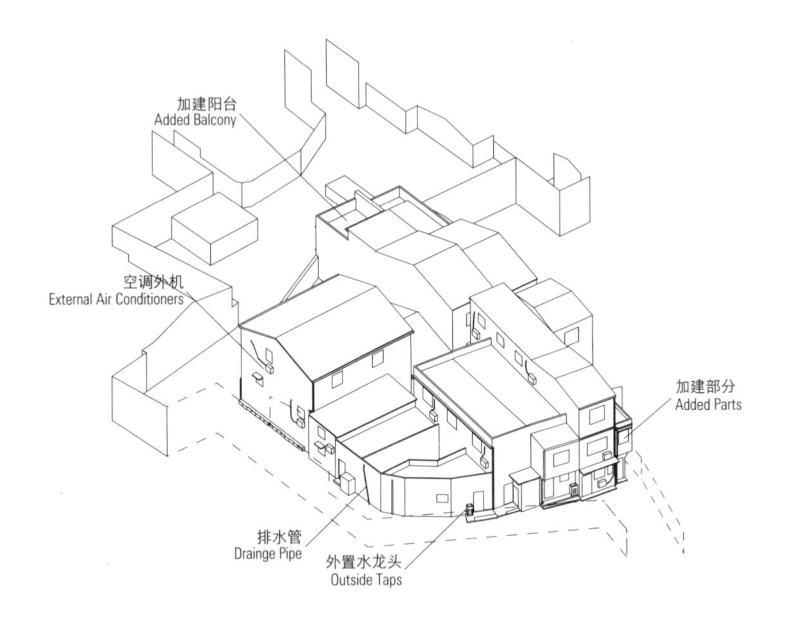

一只大象踩不到一只蚂蚁，城市规划也控制不到弯巷。从老里弄的住宅自我更新发展开始，弯巷综合体在生长。空调与排水管对于弯巷综合体，就好比电梯与钢结构对于摩天大楼一样重要。住户根据自身的现实需要改造或者配置进新的功能与设备。此外，弯巷综合体里的人群自觉团结成一个整体，对外保持一定的警觉性。在调研过程中，调研员曾遭到怀疑与拒绝。

As an elephant isn't able to step on an ant, hardly can city planning get control of Curved Alley. From the start of the self-renovation of the old Lilong housing, Curved Alley has been germinating. For it, air conditioners and drainpipes are as important as lifts and steel structures for a skyscraper. The inhabitants change the old functions and facilities by bringing in new ones according to their own needs. What's more, all the inhabitants within Curved Alley have consciously united together to keep vigilance of the outside. During the survey, our investigators have encountered their suspicion and refusal.

场所：虹口区中州路与虬江支路交接处
功能：居住
Site: intersection of Zhongzhou Road and Qiujiang Branch Road, Hongkou District
Function: residence

45 补丁之家 Bricolage House

补丁之家位于一个热闹而嘈杂的市场街道中，周围一片都是由小楼房堆积起来的低层高密度区域。补丁之家大约建于 20 世纪 60 年代，房子共 3 层：首层是店铺；二层为储物房；三层依旧有人居住，保留相对完整。外墙面上使用砖块、抹灰、泥土等多种建材及多种砌筑方式，顶楼的窗户木窗框与铝合金窗框混杂，屋顶则使用废旧的生锈钢板。

Bricolage House is located in a busy and noisy market street, surrounded by several high-density areas of compacted low-rise buildings. Bricolage House was erected around 1960s and contains three floors. On the ground floor there is a shop, while the second floor is used for storage, with the holder living on the third floor. Various construction materials (such as brick, plaster and mud) are used by various building methods on the external wall. Wooden and aluminum frames are combined in the windows on the top floor, and old rusty steel plates are employed for the roof.

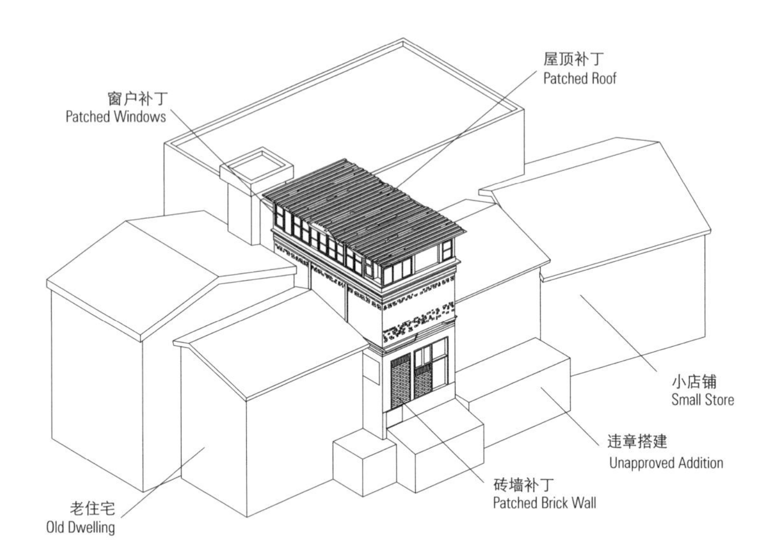

补丁之家的特别之处在于，由于不断地修补而看起来像一幅拼贴画。处处可见的民间自主修葺痕迹——屋顶的补丁、外墙新砖的补丁、窗户的补丁——以不同的材料和方法与原有的沧桑部分并置在一起，衍生出一种源自本能而非有意地形态。它们更是一种见证：时间的见证，建筑和人都在努力生存的见证。

Bricolage House is special for its continuously mending and repatching, which makes it look like a collaged painting. The self-help patching traces everywhere — the patched roof, patched brick wall and patched windows — are juxtaposed with the original part by different materials and methods. They have developed a kind of shape coming from instinct instead of intention. They are more like a witness: witness of time, and witness of the effort of both a building and humans to survive.

场所：卢湾区建国中路 171 号
功能：居住
Site: No.171 Middle Jianguo Road, Luwan District
Function: residence

46 泰康平台 Taikang Terrace

泰康平台原先是街道办拉丝模具厂的老厂房，保留了原先的建筑结构和大部分砖墙。在高密度的居民区内，它采取了向天空开放的策略，将各个屋顶积极地利用起来用作休闲、办公和艺术创作。虽然现在的泰康平台远没有田子坊那么著名，但它应当属于几近成名的创意园区副中心，吸引的主要不是慕名而来的人群，而是偏爱隐居玩乐的人士。

Taikang Terrace was formerly an old draw bench mold factory. It has retained the original structures and most of the brick walls. Enclosed by the surrounding high-density residences, Taikang Terrace turns to the sky, making the most of all the roofs for entertainment, offices or art creation. Though now Taikang Terrace is far less famous than Tianzifang, it should be regarded as a sub-center of creative parks. It mainly attracts people who seek a life in seclusion instead of those who come to merely admire.

24h
ICBC
171

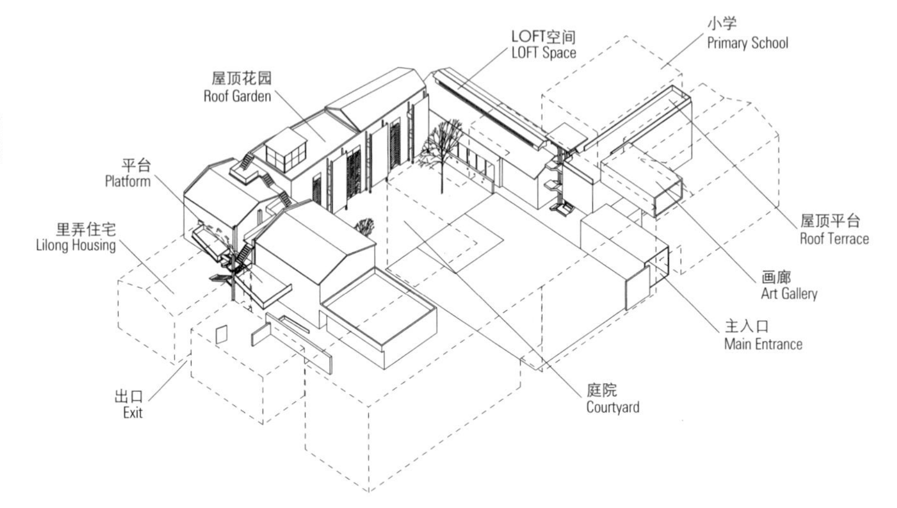

与田子坊的“平原式”格局相比，泰康平台更像是复合式的“高原地区”。它在剖面上有着层层深入的空间，顺着老式木质楼梯悠哉向上，由一系列不同标高的平台构成的空间在连绵的里弄住宅里创造出一片异质的孤岛。这些平台是本案例的最大特点，它们与居民楼仅相隔咫尺。当一位白领在平台上享受咖啡的时候，旁边一米外的住宅内居民正在洗脸刷牙。公共性与私密性的高度叠加缝合了改造后的新空间实体与原有的老房子。

Compared with the "flat plain" layout in Tianzifang, Taikang Terrace seems more like a compound "plateau section", with layers of in-depth spaces. Ascending along the vintage wooden staircases, you will find a space made of a series of terraces on different levels, acting as an islet within the continuous ordinary Lilong housing. This case is characterized by such terraces, which are merely a few inches beside the local houses. When a white-collar pedestrian is having coffee on the terrace, a local inhabitant may be washing his face just one meter aside. The extreme overlap of public and private spaces unites the newly-renovated spaces and the original old houses.

场所：杨浦区杨树浦路 500 号
功能：居住
Site: No.500 Yangshupu Road, Yangpu District
Function: residence

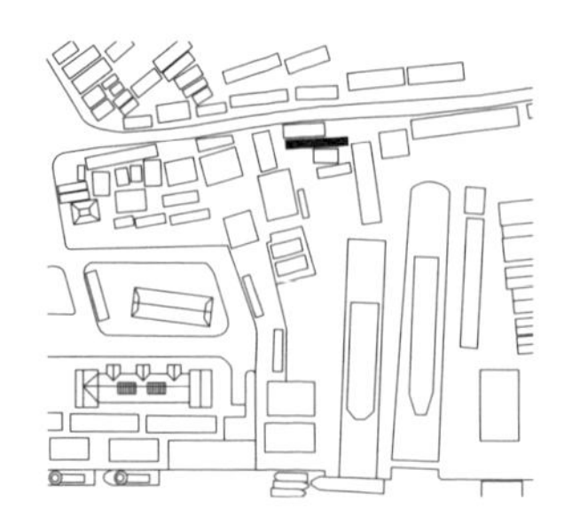

47 挡灰公寓
Dust-blocking Apartment

南面是尘土飞扬的上海船舶公司机械加工中心，北面是居住小区。为避免灰尘蔓延到居住区，两者之间竖立起了一片钢框架的挡板。与此同时，在钢框架的底部三层，插入了箱体单元供工人居住。每一个单元内摆放四张上下铺的床，窗边一张桌子，摆着一台电视机。走廊上非常阴暗，房间门口都堆放船厂统一的工鞋，并拉绳晾晒衣服。钢框架的上部三层还空置着，只看得到钉上去的铁皮。

Located on the south side of this case is the dust blowing machining center of Shanghai Shipping Firm, while on the north side there is a residential quarter. A dam-board made of a steel frame is set up between them in order to block the dust. At the same time, living units for the workers are inserted into the lower three floors of the steel frame. In each of the units there are four bunkbeds and a table by the window with a television. The corridor is very gloomy. The workers place their boots at the door and hang up their clothing on the clothesline. The upper three floors are still vacant, with only iron plates attached to the steel frame.

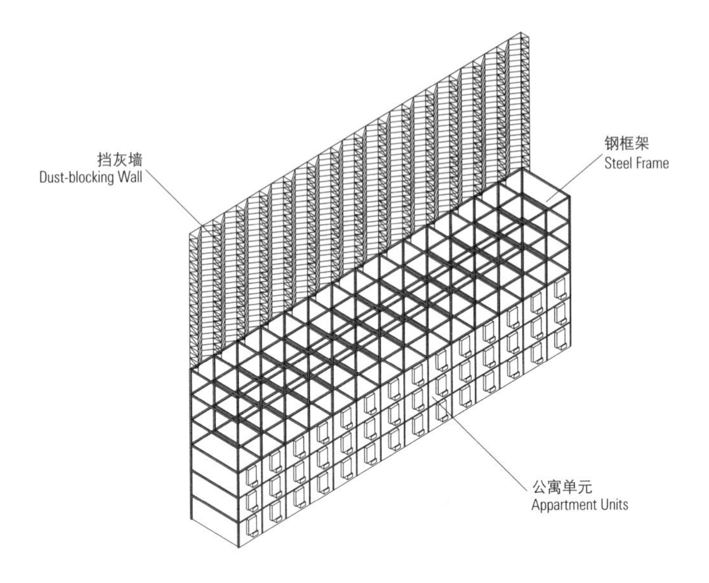

挡灰公寓由于年久失修而导致外观破败不堪，使人感觉这是个伫立在船厂与居住区之间的钢铁怪物。它自身被植入的居住功能部分削减了外部形态带来的消极性。挡灰的功能与日常的居住功能共生一体——在路人看来，这是个钢铁怪物；但对船厂工人们来说，这是一个温暖的家。

Worn down for years without repair, Dust-blocking Apartment appears in ruins, making itself as a steel monster between the shipping firm and the residential quarter. Its implanted residential function partly remits the passiveness brought by its external shape. It is a dust-blocking board as well as a house for everyday life, with these two functions coexisting. In the eye of a passer-by it seems like a steel monster, while for the workers it is a large warm house.

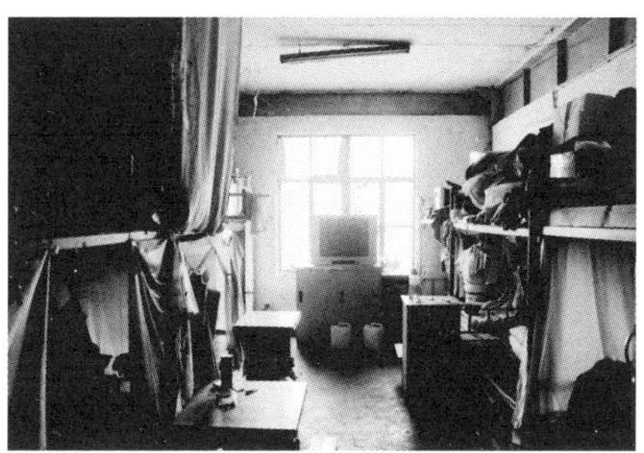

场所：浦东新区凌兆新村地铁站附近
功能：居住
Site: near Lingzhao New Village Metro Station, Pudong New Area
Function: residence

48 集装箱公寓 Container Apartment

在上海市郊，七零八落地堆砌着许多废弃的集装箱，它们成为一些低收入人群的容身之地，房东收取每个集装箱每月 500 元的租金。这里住着不同的家庭，简陋的摆设中挤满了与空间面积不相符的人口。住户们自发地给这种集装箱的家装上玻璃门和铝合金窗，并从旁边的工地上拉来电线和水管线，再购买一些简单家具悉心安置到集装箱内。这些基本满足了他们的生活需求。

In this suburb of Shanghai there are many deserted containers, which have become the inhabitance of low-income groups. The owner of these "apartments" rents them at a price of 500 yuan per month for each container, with different families living in an extra high-density community. The tenants have installed glass doors and aluminum window frames for their container apartments. The tenants even draw electricity and water from the construction site nearby. They also have bought some simple furniture and carefully placed them inside the container. All of these have met their demand of living.

屋顶阳台
绿化工作室
电话:15800567243
苗圃
电话:58411262
ILSU 2006966
MLCU 224403 6
KMBU 296116
MAX GR
TARE
NET
CU CAP

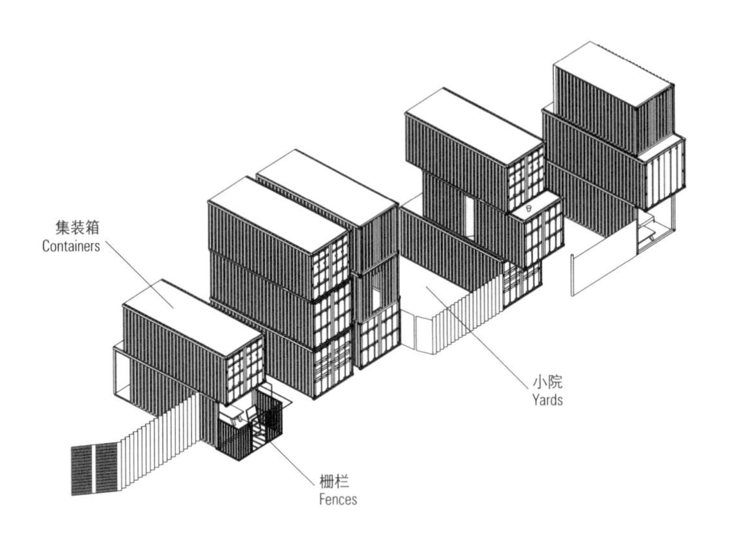

原本用于装运货物的集装箱废弃后成为消极空间，但当这些外来务工人员的居所被植入后，它们焕发出新的意义。在这些临时的家里，父亲与儿子在外面的道路上玩耍，母亲在没有窗户的集装箱中整理衣物。虽然这种“蜗居式”的公寓在质量安全、消防、通风采光等方面都有诸多问题，但对这些外来务工人员来说，这已经是让他们生存于这个大都市中的最大慰藉。

The containers become very passive when they are deserted from transporting cargos, however, they are rejuvenated after the apartments of these migrant workers are implanted. Within these temporary homes, a father is playing with his son on the road and a mother is folding clothes inside this container without windows. Though there remain a great many problems in safety, ventilation and fire protection in such a "Snail Shell Dwelling", these container apartments have provided the greatest comfort for the migrant workers in this metropolis.

COSCO

场所：浦东区崮山路 311 号
功能：菜场 + 停车场 + 小公园 + 浴场 + 饭店 + 小贩住所
Site: No.311 Gushan Road, Pudong District
Function: food market + parking lot + small park + baths + restaurant + housing of vendors

49 菜场河 River of Market

泾东菜场的所在地在是荻柴浜，20 世纪 90 年代末，为了满足附近居民买菜的需求，政府将荻柴浜这一段河道铺上盖板后建造了泾东菜场。菜场最前段的街面是小饭馆，紧接着是杂货铺，中段的店铺售卖蔬菜、肉、酱料等，尾段是水产店，最尾端是垃圾处理。西侧的小路成为停车场及流动菜摊摆卖的空间，北边西侧两层高的房子部分用于小贩住处。

The autonym of this case is Jingdong Food Market. Originally it was Dichai Creek that flowed here, however, in the late 1990s, the creek was paved and the market was built upon it in order to facilitate the public. Located at the front is a small restaurant and several grocery stores. Shops selling vegetables, meat and sauce are in the middle, and at the end there are seafood shops, and a garbage disposal site following. The alley on the west becomes a space for parking and mobile vendors, and part of the 2-story houses on the north are for the housing of these vendors.

东集贸市场

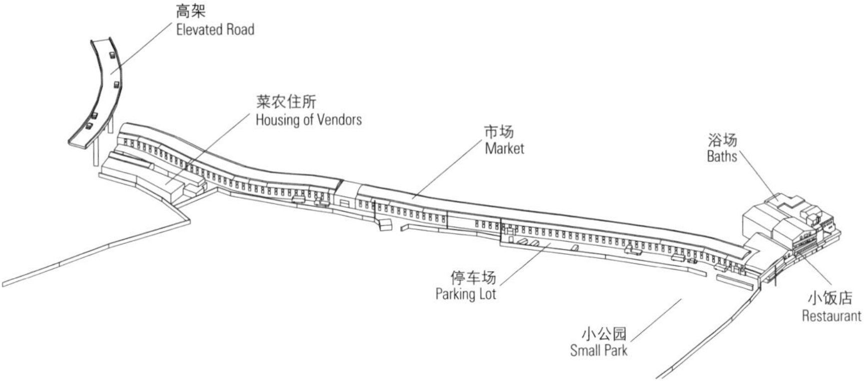

高架
Elevated Road
菜农住所
Housing of Vendors
市场
Market
浴场
Baths
停车场
Parking Lot
小饭店
Restaurant
小公园
Small Park

“菜场河”是大都会中生态景观被人造物替代的典型案例。市场的形态来源于底下的河，河道功能被置换为贩卖功能。增加出来的面积被安置上一个公共性很强的菜场后，填补了小区之间的原有空隙，大大拓展了一条河或者一条路所能发挥的城市功能，带给这片边界区域强大的活力。与此同时，经过疏浚的荻柴浜也重焕了生机。

"River of Market" is a typical case of ecology landscape being replaced by a man-made construction in a metropolis. The shape of the market is derived from the creek below it and the function has been turned from a canal to selling. On the created area is a market with strong publicity, which has filled in the interspaces of the residential quarters. The urban function of this market has brought energy to this boundary region, thus exceeding a mere creek or road. Meanwhile, Dichai Creek has been rejuvenated after being dredged and cleaned up.

大胡子龙虾王
浴场

场所：杨浦区四平路、黄兴路、翔殷路、淞沪路、邯郸路交接处
功能：交通＋购物＋休闲娱乐
Site: intersection of Siping Road, Huangxing Road, Xiangyin Road, Songhu Road and Handan Road, Yangpu Disitirct
Function: transportation + shopping + entertainment

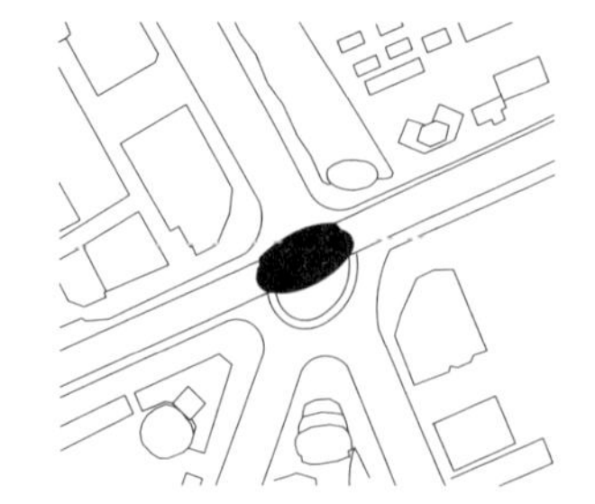

50 城市巨蛋
Giant Urban Egg

五角场因五条大路的交汇而得名，作为上海市的四个副中心之一，承担着复杂的城市功能。巨型“彩蛋”，即五角场环岛工程，是申城高架道路上首座景观工程。它长 106.8 米，高 15.8 米，重约 620 吨，并采用了 LED 照明等高科技手段。巨蛋上的彩灯每隔 10 秒变换一次颜色，共可变换红、橙、黄、绿、青、蓝、紫七种颜色，从而创造出一个迷幻璀璨的夜上海。

Wujiaochang gets its name from the intersecting of five avenues. As one of the four sub-centers in Shanghai, it has vastly complicated urban functions on its shoulders. The giant "Painted Egg", nickname for the Wujiaochang Roundabout Project, is the first landscape project on the elevated roads in Shanghai. It is 106.8 meters long, 15.8 meters high, with a total weight of 620 tons, and it has adopted high technologies such as LED lighting. The lights shining on this giant egg can change their colors from red, orange, yellow, green, cyan, blue and purple every 10 seconds, thus creating a bright and illusive Shanghai night scene.

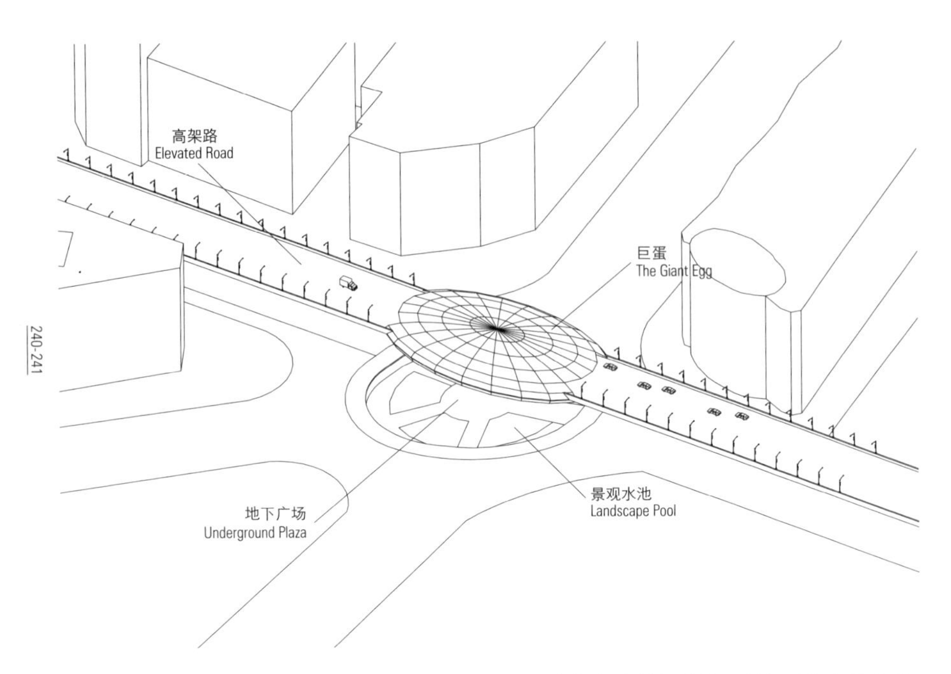

城市巨蛋以其巨大的造型、浮夸的灯光效果成为一道耀眼的人造奇观。这颗闪耀的城市巨蛋成为曾经的民国大上海城市中心的当代地标。在它下方是交通、购物、娱乐等多种功能的集散点。城市的巨型基础设施披上标志性的外衣，它一方面表征着大都会的时髦潮流，另一方面以超大型商业综合体吸引着消费者。

Giant Urban Egg is a dazzling man-made landscape for its gigantic shape and spectacular lighting. This shining egg has become the contemporary landmark at center of the former Great Shanghai Project during the reign of the Republic of China. Beneath this egg is the gathering and scattering of various functions such as transportation, shopping and entertainment. The giant urban infrastructure is dressed with an iconic overcoat, on one hand manifesting the modern fashion of a metropolis, while on the other hand attracting consumers through an ultra-large commercial complex.

YouYiCheng
百联又一城购物中心

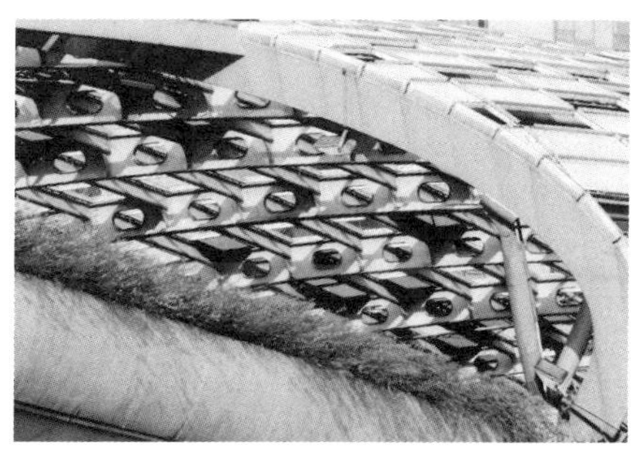

场所：黄埔区金陵东路与福建南路交接处
功能：居住
Site: intersection of East Jinling Road and South Fujian Road, Huangpu District
Function: residence

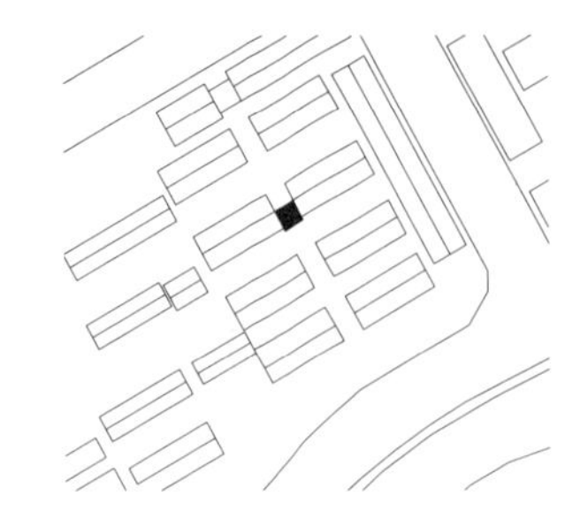

51 塔居 Tower Residence

塔居坐落于在两条旧式里弄的缝隙中，面宽 4 米，进深 5.8 米，高约 21 米，一部公共楼梯串联了 4 户独立家庭。为解决住房不足的问题，修房队（隶属房管所）将这座原本废弃的水塔改造成独立的迷你居民楼。塔居的底层由 8 根立柱架空，作为小区的主要通道，仅留出非常狭窄的楼梯通道空间。建筑自下往上逐渐朝外出挑，第二层到第五层的住户户内面积逐渐递增。每一户都在接近 4 米的室内净高中搭建了夹层。

This tower residence is located in the narrow gap between two old Lilong housings. It is 4 meters wide, 5.8 meters deep, about 21 meters high, with a staircase linking four isolated families. A local organization called House Constructing Group renovated an originally deserted water tower into this mini residence. Supported by 8 columns, the tower is elevated above the ground floor, leaving a narrow passage. The tower stretches out bit by bit from bottom up, thus the floor area multiplies from the first floor to the fourth. Each home has an indoor net height of almost 4 meters, and they all have constructed an interlayer.

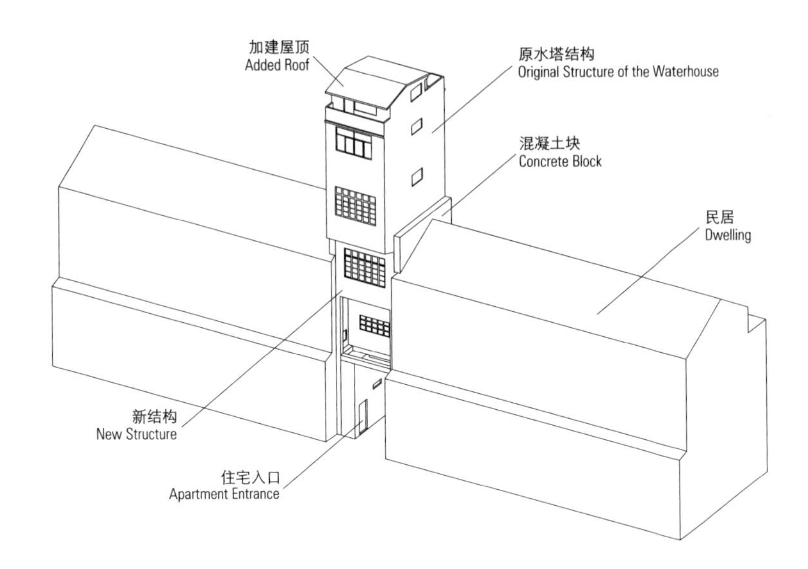

在当时追求实用的导向下，塔居这个特殊的时代产物成为以官方身份搭建的非正式建筑。在自己独立的小世界中，它以极小的面积取得了极大的居住空间，以及紧密的邻里交往。人们在居住的安全感与交往的熟悉感中慢慢与塔居融为一体。修房队当年的权宜之计所获得的意义一直延续到当代。汇入日常建筑中的民间智慧焕发出巧妙的空间组合方式，随着时间推移，与人的栖居深深叠合到一起。

As a product of the specific policy in a specific age, Tower Residence became an irregular building of an official status in the orientation of pursuing utility. In its isolated world, Tower Residence has maximized the living space and neighborhood communication while maintaining a minimum of floor areas. The inhabitants have become integrated with Tower Residence with a sense of safety and familiarity of living and contacting. The makeshift of House Constructing Group has remained meaningful today. The folk wisdom embedded in this everyday architecture has resulted in brilliant spatial compound modes, which deeply unite the dwelling of humans throughout time.

场所：虹口区瑞虹路与沙虹路交接处
功能：居住
Site: intersection of Ruihong Road and Shahong Road, Hongkou District
Function: residence

52 鸽舍 House of Pigeons

鸽舍所在的场地上在新中国成立前是棚屋，后来盖起了这幢砖混房屋。顶部用木板加建出大小不一的笼子，用来饲养鸽子。木笼子还借助木头斜撑固定在外墙上。主人是一位近 60 岁的无业居民，他与近 90 岁的老母亲住在楼下。

Before the founding of PRC, there was a shed on the site that later became this brick-cement building. On the top, various-sized cages made of wooden planks were added for feeding pigeons. These wooden cages are fixed to the external wall by wooden strut. The householder is a jobless man who is nearly 60 years old. He lives on the lower floors with his old mother, who has an age of about 90.

241

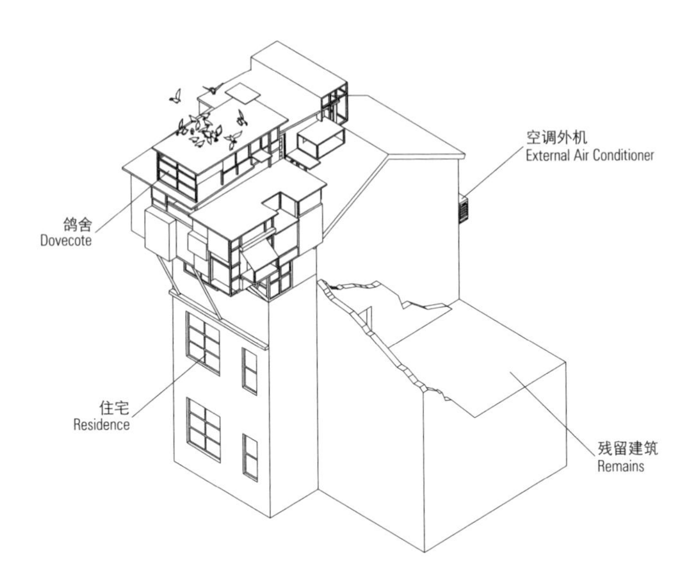

在上海，老房子顶部加建出鸽舍的例子并非罕见，但为每一只鸽子都搭建出专门空间的恐怕只有此例。主人对生活的热爱促使他为鸽子搭建了如此“豪华”的居所。他告诉调研者，这片场地位于政府拆迁范围之内，几年之间这栋房子就将消失。

It is not rare in Shanghai to find a case of an added pigeon house, but this case is unique because each pigeon is provided a specific space. It is the love of life that prompts the householder to add such "luxurious" houses for his pigeons. He told us that this house might disappear in a few years, because it is in the demolition scope defined by government.

场所：闵行区莘建东路
功能：火车 + 地铁 + 公交 + 出租车 + 快餐
Site: East Xinjian Road, Minhang District
Function: railway + metro + bus + taxi + fast food

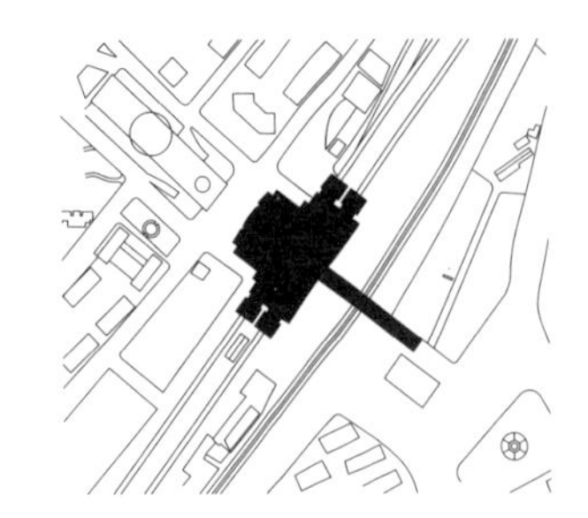

53 莘庄地铁站 Xinzhuang Metro Station

莘庄地铁站不仅是1号线的终点站，同时也是1、5号线的换乘站。近年来的改造使其原本割裂的南北地块联系更为紧密，通过一个大平台连接南北广场，实现便捷换乘。大量人流往来于此，该站成为集轨道交通、铁路、公共交通和高速、快速道路交通等多种陆上交通形式为一体的综合交通枢纽。

Xinzhuang Metro Station is not only the terminal of Metro Line 1, but also the interchange station of Line 1 and 5. The current reconstruction has integrated the originally disconnected street blocks. It uses a giant platform to link the squares at the north and south, from which many passengers will benefit. Xinzhuang Metro Station has become a large junction of various land transportations such as metro, railway, bus and expressways.

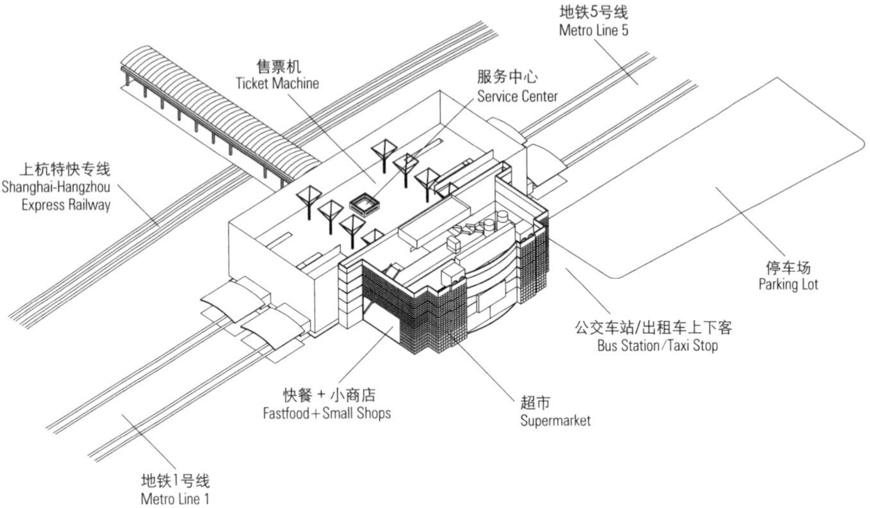

莘庄地铁站是位于上海中心城区外围的交通枢纽，是接驳上海与它周边的卫星城的中转站。上海作为当下中国综合交通集成化程度最高的城市之一，通过交通动脉的延伸为城市化过程铺平道路。此外，巨型城市基础设施也打破了一栋建筑在空间上的界限——莘庄地铁站虽是一栋单体，但它拼合起来的多种交通设施让它成为一台城市中时刻运作的机器。

Being a transportation connection at the periphery of Shanghai central city, Xinzhuang Metro Station unites Shanghai and its satellite towns. As one of the most developed city in the integration of comprehensive transportation, Shanghai has paved the way for urbanization by the extensions of such transportation arteries. Additionally, large urban facilities have broken the boundaries of architectural space. Though Xinzhuang Metro Station is a single building, it seems like an urban machine running all the time.

换乘 5 号线
出站验票
换乘 1 号线

场所：黄浦区复兴东路与巡道街交接处
功能：居住
Site: intersection of East Fuxing Road and Xundao Street, Huangpu District
Function: residence

54 小炮台 Small Barbette

“小炮台”是周围居民对这栋已建成五六年、高达 6 层的违章建筑的戏称。在一片破败拥挤的平房中，这座鹤立鸡群的小楼显得格外亮眼。一条只容一人通过的楼梯将各楼层串联起来，原本狭窄的通道在杂物的堆放下更显混乱。塞满住户的小楼在沿街面却没有任何开窗。

"Small Barbette" is the nickname given by the neighbors to this 6-floor unapproved building which has been constructed for about 5 years. It stands out within a group of run-down shanties. All of the floors are linked together by a staircase, which is so narrow that only one person can go on it. Additionally, the staircase is stacked with all sorts of sundries, thus making the space seem increasingly narrow. The slim building is filled with inhabitants, while it has no windows on the side facing the street.

南
99
巡道街
Xundao St.
15
N
S
中国移动通信
手机维修
手机厂家直销
手机贴膜
回收二手机

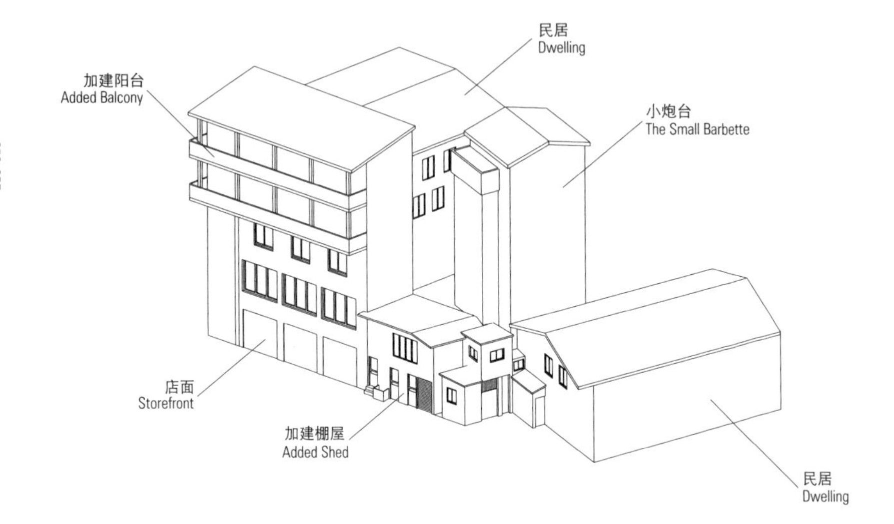

上海老城厢内的面貌有着与现代化国际大都市形象完全不同的另一面。为了在这个拥挤的区域里创造更多的居住空间，这里充斥着各式各样的搭建建筑。在各种苛刻的条件下，人们唯有通过这种方式才能在这个城市中谋得一寸安身之所。“小炮台”的存在折射出这种逼仄的生存环境。

Compared to the modern areas of Shanghai, there is a totally different face in the old quarter. Various added houses are compacted into a small area in order to provide more residential space. Only in such a way can these local inhabitants find a place for themselves and their families. "Small Barbette" reflects this cramped living environment.

南
巡道街
15
Xundao St
S
N

沿苏州河边行走：访谈塚本由晴
Walking along Suzhou Creek: an Interview with Yoshiharu Tsukamoto

江嘉玮
Jiang Jiawei

一

一月初的上海，北风凛冽。今天天阴，冷，我需要陪同来自东京工业大学的塚本由晴老师参观几个“上海制造”的案例。早就熟悉塚本老师主持的犬吠工作室（Atelier Bow-Wow）的若干项目，今天我打算除了完成导览工作外，再当面请教塚本老师一些问题。在地铁上简单梳理了这些问题，同时，我还在猜测：现在一共有五十多个案例，塚本老师会选择去看哪几个呢？

提前到达他下榻的外滩源北侧的上海大厦等待。不消一会儿，塚本老师从电梯里出来，正如我在网络上对他的印象一样，内敛而沉稳。握手过后，我连忙将地图和案例列表摊开，塚本老师笑了笑，从挎包里掏出厚厚一叠 A4 大小的文稿。原来他早就将我之前发过去的电子文稿全都打印出来并且仔细阅读了。

“我打算沿着苏州河边步行来感受这个城市。有哪栋房子你建议我们先去看的吗？”
“沿着苏州河边走……嗯……停车塔，我觉得那是个有趣的案例。”

本以为整个下午会坐着出租车四处逛，没想到他提出要沿着苏州河走。走出酒店大门时，塚本老师还跟我说，昨天他从静安区步行了两个多小时到人民广场。如此一看，我估计塚本老师一定是希望在步行的过程中挖掘到他感兴趣的内容。沿着苏州河走很明智，因为外滩沿岸很多老房子已经面目全非，而苏州河边还是有不少值得一看的老房子。沿着苏州河河畔走，能够遇到一系列的案例。比如，废墟游乐场（案例 8）、高架综合体 I 与 II（案例 12 与 32）、停车塔（案例 19）、三明治学校（案例 25）等。

我们从外白渡桥开始沿着苏州河向西行走。走不多远，遇上一群人围在路边打扑克，塚本老师很感兴趣地凑过去拍了张照。又走不远，遇上拉着满满一车箱子的三轮车夫、在门口分类垃圾的老爷爷、坐着里弄巷口聊天的老太太们，他都会耐心的逐一拍照。显然，这些都与他对行为的研究有关。苏州河边的围栏很高，他基本不靠着河边走，而是经常钻进里弄里。专业嗅觉促使他深入到最日常的建筑和生活中去找寻有趣之处。

穿过一个寻常的里弄大门，内部的日常空间展现出来。砖木结构的两层房子呈 U 字形围合，内院比较狭窄，种植有花草。在二楼两侧由木头搭建而成的走廊上，两三位住民正在清洗日用品或者在做饭。内院上空横跨很多根长长的竹竿，有些住户将衣物凌乱地晾晒在上面。上二楼的楼梯由木头搭成，扶手上有简单的雕花，虽然老旧但仍然精致。楼梯旁边是两个公共的洗手台，洗手台下的青石板上长了青苔，排水孔周围黑黝黝一片。这是上海里弄内很寻常的一个场景，而塚本老师不仅很喜欢这个场景，而且对被岁月冲刷过的老物件非常赞赏。他建议我们将它作为一个新的案例添加进展览里。他还很喜欢那些户门前悬挂着的鸟笼。

走出里弄的内界面，我们继续走在它的外界面上。来往的车辆很多，而塚本老师经常在街道的两侧来回穿行。他很注意观察沿街立面的一些细节，比如说，他会提醒我去留意一栋位于十字路口转角位置房子的窗子，两侧窗子都平着外墙，而这个转角的窗子却将窗台凸出来。对于两栋并排建筑之间的共用山墙，他提醒说这可能会与房屋产权有关。他还对一些窗子很小、甚至没有窗子的建筑（一般是通风塔之类的机械式建筑）也很有兴趣，这多少都与他那本《窗景》（*Windowscape*）一书有点关联。相反，他对新建的大型建筑一般不大感冒，不过，他对某些功能并置的建筑很有兴趣。比如说，来到上海笑天地黄浦剧场时，他发现在剧场主入口旁边的首层用作小商场，就很有兴致地钻了进去。

我们接着走。当经过一个位于某栋建筑首层的菜市场时，塚本老师忽然走了进去。里面是中国式菜市场最普通的场景——各式各样的摊位、夹杂腐烂气息的腥臭味、讨价还价的对话。塚本老师不仅对这个空间以及内部的行为感兴趣，他还对那些贩卖的商品感兴趣。他指着一种长得像向日葵的蔬菜问我：“你知道这是什么蔬菜吗？”我也只得耸耸肩回答不知道。然后，我问他这样的菜市场在东京普不普遍，令我感到意外的是，他回答在东京没有这样的市场。“家庭主妇从超市里买菜。”哦，原来东京已经相当普及大型超市了。而这种尚且处于原始状态的菜市场，只能以批发市场的形式出现，不会再被作为零售了。

终于来到停车塔。他远远望见这栋外立面使用粉色铝板的房子时就基本都熟悉情况了，所以等到经过车塔底下时，他直接大步地往前走过去。我提醒他可以绕到车塔的另一端看看，他回答说不是很必要。走出不到十米远，他忽然想起什么，带着我又绕回去。原来他发现车塔背后有一个巨型的长条状建筑。这栋建筑整整有接近400米长，中间开有三个可通车的门洞，两侧还各有一个地铁出入口。

我们再往前行，又翻越几个街区后，天开始全黑了。于是我们搭乘出租车来到 M50 创意园区，这是一个艺术创意中心与艺术展区。陪他逛过几家艺廊之后，最后我们回到创意园区门口的一家咖啡厅喝咖啡。坐下来后，他一边喝咖啡一边和我翻看选好的案例，讲了一些他的初步评价。他喝咖啡的速度很快，我的这杯摩卡还剩三分之二多，他那里的已经全喝光了。

在昏黄的灯光下，我向塚本老师展开了一段小访谈，话题集中在今天的行走路线与一些与当代城市状况相关的理论。

二

江嘉玮（以下简称“江”）：塚本教授，现在我想请教您一些关于“空间的表象（再现）”与“表象（再现）的空间”的问题。前者对应着使用者，而后者与建筑师或者设计师有关。对吗?

塚本由晴（以下简称“塚本”）：不，不！我认为它们两个恰好相反。

江：您能详细解释一下当中的区别吗?

塚本：当然可以。这些观念最初来自法国哲学家亨利·列斐伏尔。你知道他吧?

江：知道的，《空间的生产》。

塚本：正是。再现意味着你需要使用制图和模型来相对抽象地展现出空间。这个过程需要经历思考，因此空间的再现正是建筑师要面对的问题。反过来，再现的空间是设计的结果，因而它是对于使用者而言的。使用者常常不懂怎样设计，但他们肯定都会对空间有某种情感。

（这两对概念来自塚本由晴在其著书 *Echo of Space / Space of Echo* 中讨论的话题。他坦言概念直接来自列斐伏尔。参见 *The Production of Space*, translated by Donald Nicholson-Smith，P38）

塚本：你看，在很多例子里，空间直接被使用者创造出来，不需要很多专业设计。这些就是所谓的没有建筑师的建筑。这种现象说明实际上绘图和建造过程是可以分开的。西方历史上，在文艺复兴之前，鲜有建筑留有建筑师的名字。

江：嗯。文艺复兴期间的状况是，很多大教堂都由著名的建筑师建造或者改造，比如说阿尔伯蒂。我们很少得到任何关于中世纪建筑师的信息。

塚本：而从文艺复兴以来，制图开始从常规的建造过程中分离出来并且成为一种专业化的设计途径。从那开始，空间的再现与再现的空间转变成为两个对立面。在工业革命前后，这两者的分离很严重，人们试图通过引入第三者来调解当中的矛盾。这个第三者就是空间的实践。请注意，这个单词是“s-p-a-t-i-a-l”，不是“s-p-e-c-i-a-l”。

（毫无疑问，塚本老师对这两对概念的阐释彻底地来自列斐伏尔。列斐伏尔在《空间的生产》理论中构建出空间的实践、空间的再现、再现的空间这样一对三元关系。由此，列斐伏尔在马克思二元论的基础上加入了第三元，目的在于化解二元对立关系。在接下来的访谈中将会看到，塚本老师对列斐伏尔的继承，并非直接引用“空间的再现”、“再现的空间”这样的词汇，而在于同样依靠第三元来缝合二元的间隙。）

江：您从列斐伏尔那里引入空间的实践，对吧?

塚本：正是。空间实践的存在帮助将前面两者结合起来。空间的再现与再现的空间通过空间的实践互相调换并且融合。这是缩窄建筑师与使用者之间鸿沟的途径。还有，这些都与一个德语单词“扬弃”有些关系。

江：德语单词？……能再说一遍吗？

塚本：Aufheben（扬弃）。列斐伏尔在他的理论中引用的一个德语词。它来自黑格尔与卡尔·马克思。

江：哈，我想起来了！马克思的确说过所谓的“扬弃哲学”。

（塚本老师点出了空间的实践在三元关系中发挥的作用。同时，他提到 aufheben 这个德语词，显然，aufheben 充当的就是这个第三元的角色。aufheben 在德语中，可分前缀 auf- 有“在……之上”、“向上”的意思；heben 是动词，它同时含有“举起、提升”与“去除、摒弃”两者意思。换言之，aufheben 这个词本身就包含有两种矛盾、对立的含义。在马克思著作在中国早期的翻译中，aufheben 很多时候仅仅被保留其中“摒弃”的含义，而马克思批判中的“Die Aufgehobene Philosophie”也一度在长时间内被误译为“消灭哲学”。实际上，aufheben 的准确译法应该是“扬弃”。一扬一弃，将两个意思的都包含了。它大概带有点我们常说的“去伪存真”的含义，简单说，aufheben 需要很辩证地对待事物。那么，回到列斐伏尔，他的确希望通过“扬弃”来发挥空间的实践在三元关系中的积极作用。）

江：您刚才已经提到空间的实践。这让我联想起 20 世纪 20 年代阿道夫·路斯发展出来的“空间体量规划”设计策略。它正是空间的实践的一种，您赞同否？

塚本：在一定程度上我赞同。“空间体量规划”让我们知道怎样设计空间。但对我而言，这还不够，因为它至多是一个工具。不要以为工具能够解决所有问题。知觉是空间实践的一个非常重要的方面。它将建筑师与使用者联系起来。勒·柯布西耶的新建筑五点同样非常伟大，但它还是不够。空间的实践必须将工具和感知结合起来。

（列斐伏尔同样提到空间中的感知，这一点上深刻影响塚本老师。而塚本老师所理解的感知，不仅仅介乎人与物这两个层面，同样介于建筑师与使用者两个层面。空间中的感知对于他来说，是建筑师与使用者不断发生着分离的现状的缝合剂。）

江：晓得了。那就是“空间的回响／回响的空间”的原初概念。据我得悉，犬吠工作室最近出版了两本有趣的书，《东京制造》和《窗景》。在两本书中，您都对普通的事情有着非常独特的视角。您在我们日常生活的建筑中发掘到了许多深刻的内涵。请问你能否简单谈谈这两本书，并且讲讲在书中您如何考虑空间的再现与再现的空间？

塚本：好的。你看，在两本书中，我们都展示了许多图解。尽管这些图解不是由我们设计出来的，但它们正是空间的再现。比如说，东京的都市空间。相反地，书里面的照片，与这些普通建筑它们自身一样，恰恰展示着再现出来的空间。这些空间就是它们原本的设计，无论优劣。在这两本书中，我们尝试同时展示出建筑师怎样看待、思考和绘制建筑，以及使用者的日常生活。换句话说，《东京制造》和《窗景》既是概念化的，又是描述性的。我不仅对人群的行为感兴趣，我更对社会和建筑物的行为感兴趣。

江：好的，很感谢，获益良多。顺便问问，我想知道，除了列斐伏尔，您还推荐哪位哲学家的书呢？

塚本：哈哈，我总会让我的学生尽量多地阅读。读读米歇尔福柯吧。我认为他的《词与物》与《监狱的诞生》都挺棒。

江：晓得了。再次感谢！

I

Chilly wind blows during January in Shanghai. It is cloudy and cold today, and I had to guide professor Yoshiharu Tsukamoto from Tokyo Institute of Technology to visit several cases of "Made in Shanghai". Being familiar with some projects by Atelier Bow-Wow, I would like to ask professor some other questions besides being a guide. A question emerged when I was taking the metro: there are totally more than 50 cases now, so which is Prof. Tsukamoto going to visit?

I arrived at Shanghai Broadway Mansion where he was accommodated in advance. Just a moment later, Prof.Tsukamoto came out of the lift. He was almost as what I had thought of him before, reserved and calm. After shaking hands, I promptly unfolded the map and list of cases. Prof. Tsukamoto gave me a smile and brought out a pad of paper from his bag. He had already printed out and carefully read all the files I had sent him before.

"I would like to walk along the Suzhou Creek and feel the city by my foot. Which building do you think we can go to have a look at first?"
"Along the Suzhou Creek......ok......the parking tower, I think that is a very interesting case."

Originally I thought we would take a taxi to wander about, while then Prof. Tsukamoto insisted on walking along Suzhou Creek by foot. When getting out of the hotel, he told me that it took him about two hours to walk from Jing'an District to People Square yesterday. So I guessed that Prof. Tsukamoto must be interested in discovering something amusing by foot. Compared with the Bund which had changed a lot, it was wise to walk along Suzhou Creek because there still remained many old buildings. When walking along Suzhou Creek, we could meet several cases. For instance, Playground in the Ruins (case 8), the Complex under an Elevated Road (1) and (2) (case 12 and 32), Parking Tower (case 19), Sandwich School (case 25), etc.

From Waibaidu Bridge we started our walking westwards along Suzhou Creek. Soon we met a group of guys playing cards at the roadside, and Prof.Tsukamoto felt interested and took a photo of it. We walked on and saw some wagoners pulling a cart of boxes, a grandpa separating garbage at his door, several grandmas chatting at the entrance of the alley. Each time Prof.Tsukamoto would go close to take photos. It is obvious that all these are related to behaviors, which are always his interest point. Seldom did he walk close to the creek, for there were high fences on the shore. His professional training had urged him to dive into the most common buildings (such as Lilongs) and lives.

The everyday space within this Lilong would be presented when we got through a common portal. This two-floor wood-and-brick house had a layout of "U" shape, enclosing an inner courtyard full of flowers and grasses. In the wooden veranda on either side of the second floor, some inhabitants were washing dishes or cooking. A great many bamboo poles were placed across the courtyard, and the inhabitants put clothes on them to dry. The wooden staircase leading to the second floor had old balustrades with simple but delicate reliefs. Beside the staircase were two washing basins, beneath which there was a stone slate with moss. It was a common scene in a Shanghai Lilong. Prof.Tsukamoto not only loved such scenes, but also was fond of aged objects. He advised us to add this Lilong as a new case into the exhibition. Furthermore, he also liked the bird cages hung at the doorway very much.
After stepping out of the Lilong, we continued to walk along the external interface of it. Flows of cars

were passing by, but Prof. Tsukamoto preferred to keep crossing the street in order to get a better view. He paid attention to some details of the facade. For example, he reminded me to notice a window of a house at the intersection of two roads. This window had a sill out of the facade, while the other windows were flat to the facade. When we saw an in-between wall of two side-by-side houses, he said to me that it might be related to housing properties. What's more, he was very fond of buildings with small or even no windows (such as a mechanical building like the Ventilation Tower). More or less it had something to do with his book Windowscape. On the contrary, he had little interest in large new-built architecture. But the buildings juxtaposing several functions did interest him. For example, when he passed by Huangpu Theatre whose ground floor was used as a small shopping area, he went in with interest.

We went on walking. When we passed by a food market on the ground floor of a certain building, Prof. Tsukamoto suddenly went into it. Inside the food market was the most common scene – various stands, smells of rots, bargainings. Not only was he interested in the space as well as the inner behaviors, but also in the goods on the sale. Pointing at a kind of vegetable resembling sunflowers he asked me: "Do you know what this is?" I had to shrug and said no. Then I asked him whether such food markets were common in Tokyo. To my surprise, he told me there existed no such food markets in Tokyo. "The housewives buy their vegetables from the supermarket." I got to know that supermarkets had already become very common in Tokyo. Such a primitive food market could only be found as a wholesale market instead of a retail one.

Finally we arrived at Parking Tower. Prof.Tsukamoto had already got familiar with the building at the first sight of its elevations with pink aluminium panels, so he walked past it directly. I reminded him to turn to the other side of the tower, but he told me it was not necessary. Walking away no more than 10 meters, he suddenly remembered something and led me back. He discovered a gigantic strip building behind Car Tower. It was as long as approximately 400 meters, with three openings in the middle for cars to pass through, and with a metro entrance at either end.

Afterwards we continued our walk. Night fell after we had passed several blocks. Then we took a taxi to M50 which was a creative graden and art exhibition center. After accompanying him to visit several art galleries, we went to the cafe at the entrance of M50. Prof.Tsukamoto and I went through the cases when drinking coffee, and he gave me some of his own opinions. He quickly finished his coffee when I still had two thirds of my Mocha left.

In the dim light, I made a brief interview with Prof.Tsukamoto. Our topic focused on the way we walked today and some theories related to contemporary urban conditions.

II

Jiang Jiawei ("J" for short): Prof.Tsukamoto, now I would like to ask you a few questions about the representation of space and space of representation. The former one is for the users, while the latter is about the architects or the designers. Right?
Yoshiharu Tsukamoto ("T" for short): No, no ! I think they two are just the opposite.

J: Oh, so can you explain the differences in detail?
T: Of course I will. Originally the idea is from Henry Lefebvre, a French philosopher. Do you know him?
J: Yes, I do. The Production of Space.
T: Exactly. Represetation means that you have to use your drawings and models to show the space in a relatively abstract way.This process has to pass through your brain, so the representation of space is exactly what architects have to deal with. On the contrary, space of representation is what that results from the designer, so that it is for the users. Most of the time the users don't know how to design, but they must have a certain kind of feeling towards the space.

T: You see, in many cases, the space is directly created by the users without any professional design. They are the so-called architecture without architects. This phenomenon tells that drawings and the process of building can actually be divided. In western history, before Renaissance time, few architecture remained with the name of an architect.
J: En. During Renaissance, for example, the situation was that many cathedrals were built or reconstructed by famous architects, say, Alberti. Really we hardly know any information about the architects of the Middle Age.
T: And it was from the Renaissance time that drawings were separated from normal building

process and became a professional way of design. From then on, representation of space and space of representation had turned into two opposites. Around the industry revolution, the separation between the two opposites was so serious that people tried to reconcile the contradiction by introducing a third thing. That is spatial practice. Notice please, the word is "s-p-a-t-i-a-l", not "s-p-e-c-i-a-l".

J: You get the idea from Lefebvre to introduce spatial practice, right?
T: Exactly. The existence of spatial practice helps to combine the former two. Representation of space and space of representation are being reversed and merged by spatial practice. It is the way to narrow the gap between the architects and the users. What's more, these also have something to do with a German word called "aufheben".
J: a German word...May I beg your pardon?
T: "aufheben" , a German word Lefebvre quoted in his theory. It is from Hegel and Karl Marx.
J: Ha! I remember! Karl Marx did have said about the so-called "Die Aufgehobene Philosophie", and Hegel "Die Dialektische Aufhebung".

J: Professor, you have mentioned spatial practice just now. It reminds me of the Raumplan design strategies developed by Adolf Loos in the 1920s. Do you agree that Raumplan is just a kind of spatial practice?
T: To some extent I do agree. Raumplan tells us how to design the space. But for me, it is not enough because it is no more than a tool. Never consider that a tool can solve everything. Our perception is a very important part of spatial practice. It links both the architects and the users. Also, the five points of new architecture by Le Corbusier are excellent. But it is not enough. Spatial Practice must combine both the tools and the perception.

J: I see. So that is the primary concept of *Echo of Space/Space of Echo*. Professor, as I know, Atelier Bow-Wow has recently published two interesting books, *Made in Tokyo* and *Windowscape*. In both books, you have a very special perspective on the ordinary things. You have discovered many deep meanings from the architecture in our everyday life. Would you please talk a little bit about these books and how you think about representation of space and space of representation in them?
T: OK. You see, in both books, we have shown a lot of diagrams. Even though they are not the designs by us, but exactly they are the representation of space, for example, the urban space of Tokyo. The photographs in the books, on the contrary, along with the ordinary buildings themselves, are just showing the space of representation, the space of their original design; no matter they are good or terrible. In both book, we are trying to show how architects see, think and draw as well as how the users' everyday life is. In other words, *Made in Tokyo* and *Windowscape* are both conceptual and descriptive. I am not only fond of the behaviors of humans, but also the behaviors of the society and the buildings.

J: OK, thanks a lot, professor. I've learned a lot from this conversation. By the way, I would like to know that, besides Henry Lefebvre, which philosopher's book you would recommand us students to read?
T: Haha, I always ask my students to read more. Try to read Michel Foucault. I think his writtings of *Words and Things* and *The Birth of the Prison* are pretty good.
J: I see. Thanks again!

地图
Map

01 东方明珠广播电视塔
The Oriental Pearl TV Tower
02 上海展览中心
Shanghai Exhibition Center
03 上海商城
Shanghai Center
04 平安金融大厦
Ping'an Finance Mansion
05 山景房
Housing of Mountainscape
06 西藏大厦万怡酒店
Tibet Mansion of Marriott
07 楔形屋
Wedge-shaped Building
08 废墟游乐场
Playground in the Ruins
09 外滩美术馆
Rockbund Art Museum
10 都市巢居
Urban Nest
11 猪笼寨
Walled City
12 高架综合体 I
The Complex under an Elevated Road I
13 虹桥枢纽
Hongqiao Complex
14 龙柱
Dragon Pillar
15 通风塔
Ventilation Tower
16 南浦大桥盘旋高架
The Spiral Viaduct of Nanpu Major Bridge
17 八号桥
Bridge Eight
18 兵舰头
Warship-head-shaped House
19 停车塔
Parking Tower
20 大世界
The Great World
21 假山餐厅
Rockery Restaurant
22 静安寺
Jing'an Temple
23 外滩景观大道
The Sightseeing Avenue of the Bund
24 三角住宅
Triangular House
25 三明治学校
Sandwich School
26 四连宅
Four Neighboring Villas
27 忘忧宫
Palace Sans-souci
28 消防站上的棚屋
Shanties on the Fire Station
29 上海一九叁三
Shanghai 1933
30 高架旁的博物馆
The Museum beside an Elevated Road
31 消防瞀钟塔
Fire Station Bell Tower
32 高架综合体 II
The Complex under an Elevated Road II
33 桶中楼
Bucket Housing
34 水闸办公室
Sluice Office
35 艺术方盒
Art Box
36 水塔楼
Water Drainage Building
37 虹镇老街
Old Hongzhen Street
38 焚烧炉
An Incinerator
39 摩登假日
Modern Holiday
40 菜场旅馆
Market Hotel
41 高架小楼
Building beside the Elevated Road
42 街角楼
Corner Block
43 当代艺术博物馆
Power Station of Art
44 弯巷
Curved Alley
45 补丁之家
Bricolage House
46 泰康平台
Taikang Terrace
47 挡灰公寓
Dust-blocking Apartment
48 集装箱公寓
Container Apartment
49 菜场河
The River of Market
50 城市巨蛋
Giant Urban Egg
51 塔居
Tower Residence
52 鸽舍
House of Pigeons
53 莘庄地铁站
Xinzhuang Metro Station
54 小炮台
A Small Barbette

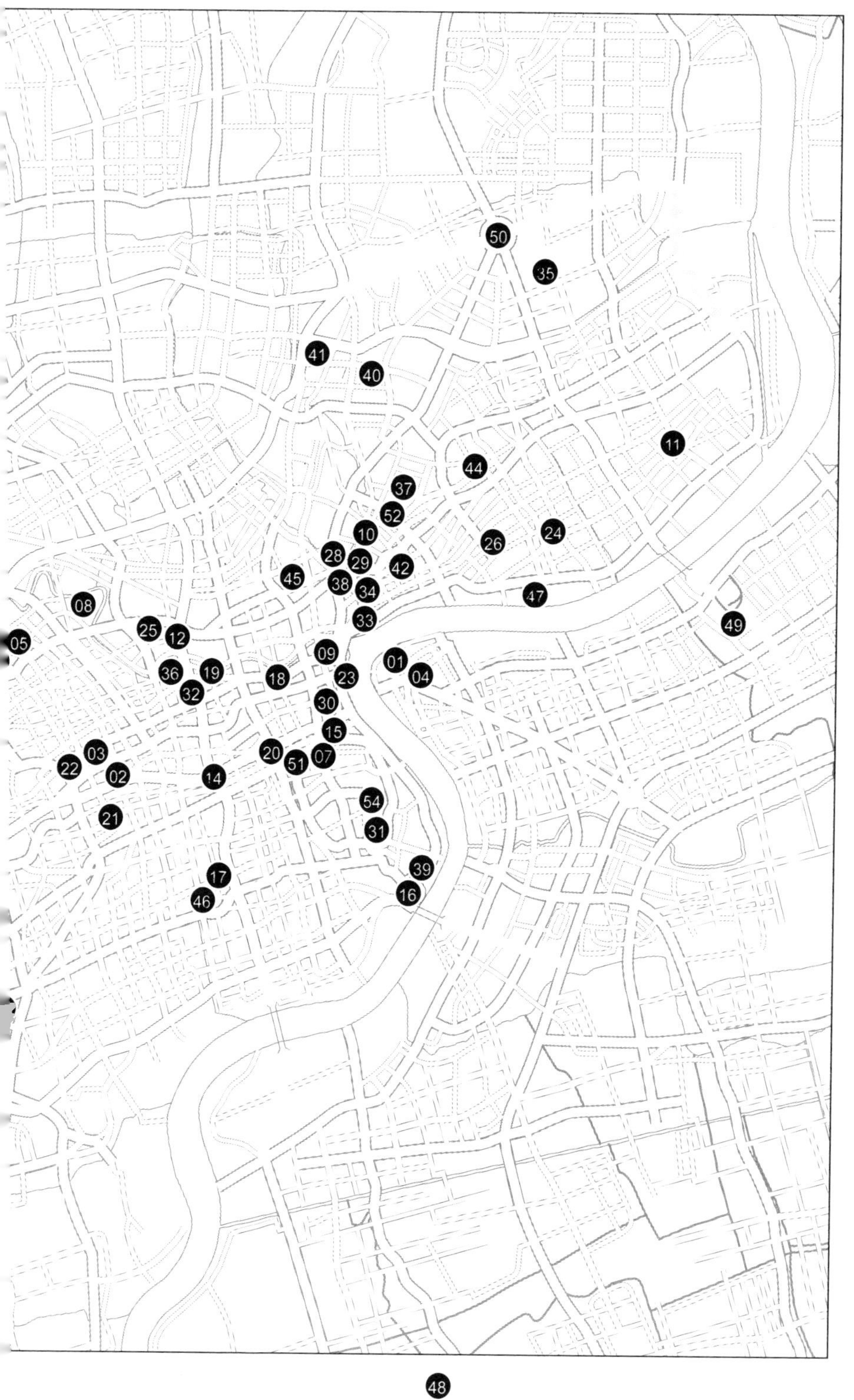
50
35
41
40
11
44
37
52
10
24
26
28
29
42
45
38
34
47
08
33
49
25
12
05
09
01
04
36
19
18
23
32
30
15
03
20
07
51
22
02
14
54
21
31
39
17
16
46

展览
Exhibition

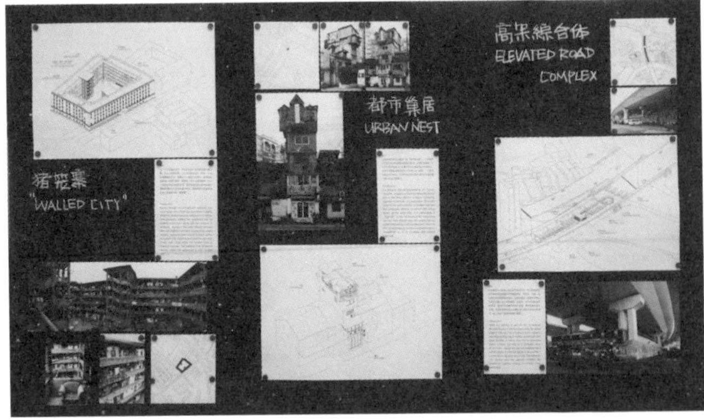

2012 年 3 月，“上海制造”研究参与在上海外滩美术馆举办的“样板屋”展览，首次与公众见面。在美术馆旁边一块长 130 米、高 2.8 米的巨型广告板上，张贴了《类推都市》研究图解。

"Made in Shanghai" project was part of the exhibition "Model Home" held in Shanghai Rockbund Art Museum in March, 2012. It was its first time to be exhibited to the public. The diagram of "Analogical City" was shown on the 130-meter-long, 2.8-meter-high billboard beside the museum.

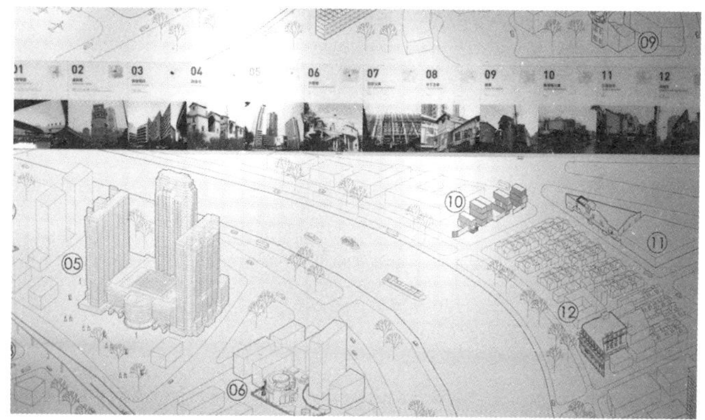

2013 年 12 月，以“城市边缘”作为主题的 2013 深圳香港城市 \ 建筑双城双年展并置了三个东亚大都会（东京、上海、香港）的都市异类空间研究。伴随着本书的出版，“上海制造”研究也开始全面走向大众。

In Decembor 2013, the Shenzhen-Hongkong Biennale of Urbanism and Architecture, which was themed "Urban Border", juxtoposed the researches of the abnormal spaces in three East Asian Metropolises (Tokyo, Shanghai, Hongkong). With the publication of this book, "Made in Shanghai" project became fully presented to the public.

项目团队
Project Team

研究指导：李翔宁

合作者：塚本由晴

研究团队：
白非、董韬、江嘉玮、李丹锋、倪正心、杨磊、杨熹、杨旭、王骁、王智聪、张鹤、张笑梅、张子岳、周渐佳

调研团队：
西蒙年科·安娜、安东尼奥、维奥拉特·布克哈特、马可·卡皮尼奥、弗兰西斯科·卡斯切拉、陈晗、陈薇伊、库尔特·吉乌索罗、莱昂纳多·西提里奥、斯蒂芬妮·克拉克、儒勒·科拉尔、蒂米泰、董晓、马克·菲奥、何英杰、伊莎贝尔·格拉尼、郭欣、华益、姜乃彬、朱莉亚、杰拉·卡斯、巴特·奇普斯、费德里克·列提吉亚、李皓、李灵凯、李鸣露、李晓旭、李鑫、林恺怡、马颖洁、提姆·曼尔、卢卡斯·马科夫斯基、桑德拉·梅、马丁·穆勒、亚当·奥德格斯、爱丽丝·彭提吉亚、曲文昕、任大任、波切特·桑迪、沈思靖、海因里希·斯巴勒、苏婷、田玉龙、王博、汪仙、魏嘉、艾丽莎·威斯坎普、吴晓帆、武筠松、徐林昊、徐蜀辰、杨洋、幺志华、于偲、张帆、钟燕、章雯、周欣

Director: Li Xiangning

In Collaboration with: Yoshiharu Tsukamoto

Research Team Members:
Bai Fei, Dong Tao, Jiang Jiawei, Li Danfeng, Ni Zhengxin, Yang Lei, Yang Xi, Yang Xu, Wang Xiao, Wang Zhicong, Zhang He, Zhang Xiaomei, Zhang Ziyue, Zhou Jianjia

Survey Team Members:
Symonenko Anna, Antonio, Violeta Burckhardt, Marco Capitanio, Francesco Cascella, Chen Han, Chen Weiyi,Kurt Chiusolo, Leonardo Citterio,Stephanie Clark, Jules Collard, Dimmitai, Dong Xiao, Michad Feo, He Yingjie, Isabel Granell, Guo Xin, Hua Yi, Jiang Naibin, Juliette, Zeller Kas, Bart Kuijpers, Federico Letizia, Li Hao, Li Lingkai, Li Minglu, Li Xiaoxu, Li Xin, lin Kaiyi, Ma Yingjie, Tim Mahn, Lukas Makovsky, Sandra May, Martin Muller, Adam Odgers, Alice Pontiggia, Qu Wenxin, Ren Daren, Portrait Santi, Sijin Shen, Heinrich Sparla, Su Ting, Tian Yulong, Wang Bo, Wang Xian, Wei Jia, Alyssa Weskamp, Wu Xiaofan, Wu Yunsong, Xu Linhao, Xu Shuchen, Yang Yang, Yao Zhihua, Yu Cai, Zhang Fan, Zhong Yan, Zhang Wen, Zhou Xin

本书为以下国家自然科学基金研究项目成果

项目名称：《我国建筑博物馆创制、博览模式及信息保存与再现技术研究》
学科代码：建筑设计与理论（E080101）
项目批准号：51078266

图片版权

“光明城”是同济大学出版社
城市、建筑、设计专业出版品牌，
由群岛工作室负责策划及出版工作，
以更新的出版理念、更敏锐的视角、更积极的态度，
回应今天中国城市、建筑与设计领域的问题。

www.luminous-city.com

图书在版编目（CIP）数据

上海制造 / 李翔宁，李丹锋，江嘉玮著．-- 上海：
同济大学出版社，2014.1（2017.7重印）
ISBN 978-7-5608-5368-0

Ⅰ．①上… Ⅱ．①李… ②李… ③江… Ⅲ．①建筑艺
术－上海市 Ⅳ．① TU-862

中国版本图书馆 CIP 数据核字（2013）第 282687 号

上海制造
MADE IN SHANGHAI

李翔宁　李丹锋　江嘉玮　著
Li Xiangning　Li Danfeng　Jiang Jiawei
合作者　塚本由晴
In Collaboration with　Yoshiharu Tsukamoto

出 品 人　支文军
策　　划　群岛工作室
责任编辑　秦　蕾　孟旭彦　　责任校对　徐春莲　　装帧设计　张　微

出版发行　同济大学出版社 www.tongjipress.com.cn
（地址：上海四平路 1239 号　邮编：200092　电话：021-65985622）
经　　销　全国各地新华书店
印　　刷　上海盛通时代印刷有限公司
开　　本　889mm×1194mm 1/32
印　　张　8.75
字　　数　235 000
版　　次　2014 年 1 月第 1 版　2017 年 7 月第 4 次印刷
书　　号　ISBN 978-7-5608-5368-0
定　　价　52.00 元